This timely volume provides a broad survey of (2+1)-dimensional quantum gravity. It emphasises the 'quantum cosmology' of closed universes and the quantum mechanics of the (2+1)-dimensional black hole. It compares and contrasts a variety of approaches, and examines what they imply for a realistic theory of quantum gravity.

General relativity in three spacetime dimensions has become a popular arena in which to explore the ramifications of quantum gravity. As a diffeomorphism-invariant theory of spacetime structure, this model shares many of the conceptual problems of realistic quantum gravity. But it is also simple enough that many programs of quantization can be carried out explicitly.

After analyzing the space of classical solutions, this book introduces some fifteen approaches to quantum gravity – from canonical quantization in York's 'extrinsic time' to Chern–Simons quantization, from the loop representation to covariant path integration to lattice methods. Relationships among quantizations are explored, as well as implications for such issues as topology change and the 'problem of time'.

This book is an invaluable resource for all graduate students and researchers working in quantum gravity.

STEVEN CARLIP received an undergraduate degree in physics from Harvard in 1975. After seven years as a printer, editor, and factory worker, he returned to school at the University of Texas, where he earned his Ph.D. in 1987. Following a stint as a postdoctoral fellow at the Institute for Advanced Study, he joined the faculty of the University of California at Davis in 1990.

Professor Carlip's main research interest is quantum gravity. He has also worked on string theory, quantum field theory, classical general relativity, and the interface between physics and topology. He has received a number of honors, including a Department of Energy Outstanding Junior Investigator award and a National Science Foundation Young Investigator award.

When he can find the time, Professor Carlip is active in progressive politics. His hobbies include hiking, world travel, playing the dulcimer, and scuba diving.

CAMBRIDGE MONOGRAPHS ON MATHEMATICAL PHYSICS

General Editors: P. V. Landshoff, D. R. Nelson, D. W. Sciama, S. Weinberg

QUANTUM GRAVITY IN 2+1 DIMENSIONS

CAMBRIDGE MONOGRAPHS ON
MATHEMATICAL PHYSICS

QUANTUM GRAVITY IN 2+1 DIMENSIONS

STEVEN CARLIP

University of California at Davis

PUBLISHED BY THE PRESS SYNDICATE OF THE UNIVERSITY OF CAMBRIDGE
The Pitt Building, Trumpington Street, Cambridge, United Kingdom

CAMBRIDGE UNIVERSITY PRESS
The Edinburgh Building, Cambridge CB2 2RU, UK
40 West 20th Street, New York NY 10011–4211, USA
477 Williamstown Road, Port Melbourne, VIC 3207, Australia
Ruiz de Alarcón 13, 28014 Madrid, Spain
Dock House, The Waterfront, Cape Town 8001, South Africa

http://www.cambridge.org

First published 1998
First paperback edition 2003

Typeset in 11pt Times [TAG]

A catalogue record for this book is available from the British Library

Library of Congress Cataloguing in Publication Data

Carlip, Steven (Steven Jonathan), 1953–
Quantum gravity in 2+1 dimensions / Steven Carlip.
p. cm.
Includes bibliographical references and index.
ISBN 0 521 56408 5 hardback
1. Quantum gravity.   2. General relativity (Physics).   3. Space and time.   I. Title
QC178.C185   1998
530.14'3–dc21   97-42893   CIP

ISBN 0 521 56408 5 hardback
ISBN 0 521 54588 9 paperback

Contents

Preface

Interest in (2+1)-dimensional gravity – general relativity in two spatial dimensions plus time – dates back at least to 1963, when Staruszkiewicz first showed that point particles in a (2+1)-dimensional spacetime could be given a simple and elegant geometrical description [243]. Over the next 20 years occasional papers on classical [77, 79] and quantum mechanical [176, 221, 142] aspects appeared, but until recently the subject remained largely a curiosity.

Two discoveries changed this. In 1984, Deser, Jackiw, and 't Hooft began a systematic investigation of the behavior of classical and quantum mechanical point sources in (2+1)-dimensional gravity [88, 89, 90, 247], showing that such systems exhibit interesting behavior both as toy models for (3+1)-dimensional quantum gravity and as realistic models of cosmic strings. Interest in this work was heightened when Gott showed that spacetimes containing a pair of cosmic strings could admit closed timelike curves [134]; (2+1)-dimensional gravity quickly became a testing ground for issues of causality violation. Then in 1988, Witten showed that (2+1)-dimensional general relativity could be rewritten as a Chern–Simons theory, permitting exact computations of topology-changing amplitudes [287, 288]. The Chern–Simons formulation had been recognized a few years earlier by Achúcarro and Townsend [1], but Witten's rediscovery came at a time that the quantum mechanical treatment of Chern–Simons theory was advancing rapidly, and connections were quickly made to topological field theories, three-manifold topology, quantum groups, and other areas under active investigation.

Together, the work on point particle scattering and the Chern–Simons formulation ignited an explosion of new research. Since the early 1980s, well over 300 papers have been published on various features of quantum gravity in 2+1 dimensions, and many others, including a large part of one book [41], have treated classical aspects. The field remains active, and few of the important questions can yet be answered with certainty. I believe, however, that we now know enough to warrant a book on quantum gravity in 2+1 dimensions.

In fact, we know enough for two books. Over the past few years, the field has split into two weakly interacting sectors, one dealing with the scattering of point particles and the other with empty space 'quantum cosmology'. This book is about the latter. To try to keep some focus, I have omitted almost all discussion of the very interesting work on point sources and matter interactions. Even with this restriction, the treatment here is incomplete and idiosyncratic: I discuss certain aspects of quantum gravity – those I understand best – in some detail, and treat others rather sketchily or not at all. Perhaps a reader will be inspired to write 'Volume II'.

I have assumed that the reader understands basic general relativity, at the level of the first chapters of Wald [274] or Track 1 of Misner, Thorne, and Wheeler [198]. Some familiarity with a physicist's version of differential topology will be helpful, although the book does not require knowledge of the intricacies of two- and three-manifold topology. I have also assumed that the reader is reasonably comfortable with quantum mechanics (canonical quantization, the Heisenberg and Schrödinger pictures, constraints, gauge invariance and gauge-fixing), and has had some exposure to quantum field theory. For the most part, however, I have not used very deep or difficult results, and when such complications were necessary, I have tried to explain them reasonably well. I end the book with three appendices, on the topology of manifolds, causal structures, and differential geometry and fiber bundles. These are not substitutes for texts, but they may help the reader through some of the more obscure sections of this work.

This book is the product of countless conversations, collaborations, arguments, and patient explanations by those who understood more than I did. Much of what I know about quantum gravity was taught to me by Bryce and Cecile DeWitt. My original interest in the (2+1)-dimensional model was inspired by Ed Witten. The idea for a book was first suggested by Gianluca Grignani and Pasquale Sodano, with whom I worked on an early version. I have learned much from my collaborators, Russell Cosgrove, Shanta de Alwis, Jack Gegenberg, Ian Kogan, Robert Mann, Jeanette Nelson, and Claudio Teitelboim. A number of mathematicians have helped me steer through the shoals of three-dimensional geometry and topology, among them William Abikoff, Scott Axelrod, Walter Carlip, William Goldman, Joel Hass, Wolfgang Lück, Geoff Mess, Alan Reid, and Scott Wolpert. I can only mention a few of the many physicists who have assisted me along the way: Arley Anderson, Abhay Ashtekar, Max Bañados, David Brown, Marc Henneaux, Akio Hosoya, Ted Jacobson,

Jorma Louko, Vincent Moncrief, Peter Peldan, Alan Steif, Jim York, and Henri Waelbroeck.

Vivian Carlip is not a physicist, but her proofreading eliminated a good number of errors. The National Science Foundation (Grant No. PHY-93-57203) and the U.S. Department of Energy (Grant No. DE-FG03-91ER40674) both supported my work, and some of the writing was done at the Aspen Center for Physics. Finally, I would like to thank Maureen La Mar and Elizabeth Beach for distracting me when I needed to be distracted.

1
Why (2+1)-dimensional gravity?

The past 25 years have witnessed remarkable growth in our understanding of fundamental physics. The Weinberg–Salam model has successfully unified electromagnetism and the weak interactions, and quantum chromodynamics (QCD) has proven to be an extraordinarily accurate model for the strong interactions. While we do not yet have a viable grand unified theory uniting the strong and electroweak interactions, such a unification no longer seems impossibly distant. At the phenomenological level, the combination of the Weinberg–Salam model and QCD – the Standard Model of elementary particle physics – has been spectacularly successful, explaining experimental results ranging from particle decay rates to high energy scattering cross-sections and even predicting the properties of new elementary particles.

These successes have a common starting point, perturbative quantum field theory. Alone among our theories of fundamental physics, general relativity stands outside this framework. Attempts to reconcile quantum theory and general relativity date back to the 1930s, but despite decades of hard work, no one has yet succeeded in formulating a complete, self-consistent quantum theory of gravity. The task of quantizing general relativity remains one of the outstanding problems of theoretical physics.

The obstacles to quantizing gravity are in part technical. General relativity is a complicated nonlinear theory, and one should expect it to be more difficult than, say, electrodynamics. Moreover, viewed as an ordinary field theory, general relativity has a coupling constant $G^{1/2}$ with dimensions of an inverse mass, and standard power-counting arguments – confirmed by explicit computations – indicate that the theory is nonrenormalizable, that is, that the perturbative quantum theory involves an infinite number of undetermined coupling constants.

But the problem of finding a consistent quantum theory of gravity goes deeper. General relativity is a geometric theory of spacetime, and

1

quantizing gravity means quantizing spacetime itself. In a very basic sense, we do not know what this means. For example:

- Ordinary quantum field theory is local, but the fundamental (diffeomorphism-invariant) physical observables of quantum gravity are necessarily nonlocal;

- Ordinary quantum field theory takes causality as a fundamental postulate, but in quantum gravity the spacetime geometry, and thus the light cones and the causal structure, are themselves subject to quantum fluctuations;

- Time evolution in quantum field theory is determined by a Hamiltonian operator, but for spatially closed universes, the natural candidate for a Hamiltonian in quantum gravity is identically zero when acting on physical states;

- Quantum mechanical probabilities must add up to unity at a fixed time, but in general relativity there is no preferred time-slicing on which to normalize probabilities;

- Scattering theory requires the existence of asymptotic regions in which interactions become negligible and states can be approximated by those of free fields, but the gravitational self-coupling in general relativity never vanishes;

- Perturbative quantum field theory depends on the existence of a smooth, approximately flat spacetime background, but there is no reason to believe that the short-distance limit of quantum gravity even resembles a smooth manifold.

Faced with such problems, it is natural to look for simpler models that share the important conceptual features of general relativity while avoiding some of the computational difficulties. General relativity in 2+1 dimensions – two dimensions of space plus one of time – is one such model. As a generally covariant theory of spacetime geometry, (2+1)-dimensional gravity has the same conceptual foundation as realistic (3+1)-dimensional general relativity, and many of the fundamental issues of quantum gravity carry over to the lower dimensional setting. At the same time, however, the (2+1)-dimensional model is vastly simpler, mathematically and physically, and one can actually write down candidates for a quantum theory. With a few exceptions, (2+1)-dimensional solutions are physically quite different from those in 3+1 dimensions, and the (2+1)-dimensional model is not very helpful for understanding the dynamics of realistic quantum gravity. But for the analysis of conceptual problems – the nature of time, the construction of states and observables, the role of topology and topology change, the relationships among different approaches to quantization – the model has proven highly instructive.

Work on (2+1)-dimensional gravity dates back at least to 1963, when Staruszkiewicz first described the behavior of static solutions with point sources [243]. Work continued intermittently over the next twenty years, but the modern rebirth of the subject can be credited to the seminal work of Deser, Jackiw, 't Hooft, and Witten in the mid-1980s [88, 89, 90, 247, 287, 288]. Over the past decade, (2+1)-dimensional gravity has become an active field of research, drawing insights from general relativity, differential geometry and topology, high energy particle theory, topological field theory, and string theory. The subject is far from being completed, but this book will summarize some of the basic features as they are currently understood.

1.1 General relativity in 2+1 dimensions

The subject of this book is the theory of gravity obtained from the standard Einstein–Hilbert action,

$$I = \frac{1}{16\pi G} \int_M d^3x \sqrt{-g}\,(R - 2\Lambda) + I_{matter},$$ (1.1)

in three spacetime dimensions. (See appendix C for my conventions for Riemannian geometry.) As in 3+1 dimensions, the resulting Euler–Lagrange equations are the standard Einstein field equations

$$R_{\mu\nu} - \frac{1}{2}g_{\mu\nu}R + \Lambda g_{\mu\nu} = -8\pi G T_{\mu\nu},$$ (1.2)

with a cosmological constant Λ that I will often take to be zero. Just as in ordinary general relativity, the field equations are generally covariant; that is, they are invariant under the action of the group of diffeomorphisms of the spacetime M, which can be viewed as a 'gauge group'.

The fundamental physical difference between general relativity in 2+1 and 3+1 dimensions originates in the fact that the curvature tensor in 2+1 dimensions depends linearly on the Ricci tensor:

$$R_{\mu\nu\rho\sigma} = g_{\mu\rho}R_{\nu\sigma} + g_{\nu\sigma}R_{\mu\rho} - g_{\nu\rho}R_{\mu\sigma} - g_{\mu\sigma}R_{\nu\rho} - \frac{1}{2}(g_{\mu\rho}g_{\nu\sigma} - g_{\mu\sigma}g_{\nu\rho})R.$$ (1.3)

In particular, this means that every solution of the vacuum Einstein equations with $\Lambda = 0$ is *flat*, and that every solution with a nonvanishing cosmological constant has constant curvature. Physically, a (2+1)-dimensional spacetime has no local degrees of freedom: curvature is concentrated at the location of matter, and there are no gravitational waves. If the spacetime M is topologically trivial, there are, in fact, no gravitational degrees of freedom at all. If M has a nontrivial fundamental group, though, we shall see later that a finite number of global degrees of freedom remain, providing the classical starting point for a quantum theory.

This absence of local degrees of freedom can be verified by a simple counting argument. In n dimensions, the phase space of general relativity is characterized by a spatial metric on a constant-time hypersurface, which has $n(n-1)/2$ components, and its time derivative (or conjugate momentum), which adds another $n(n-1)/2$ degrees of freedom per spacetime point. It is well known, however, that n of the Einstein field equations are constraints on initial conditions rather than dynamical equations, and that n more degrees of freedom can be eliminated by coordinate choices. We are thus left with $n(n-1) - 2n = n(n-3)$ physical degrees of freedom per spacetime point.

If $n = 4$, this gives the four phase space degrees of freedom of ordinary general relativity, two gravitational wave polarizations and their conjugate momenta. If $n = 3$, on the other hand, there are no field degrees of freedom: up to a finite number of possible global degrees of freedom, the geometry is completely determined by the constraints.

Now, a theory of gravity with no propagating degrees of freedom might be expected to have a rather unusual Newtonian limit. This is indeed the case: general relativity in 2+1 dimensions has a Newtonian limit in which there is no force between static point masses. To see this, let us write the metric as

$$g_{\mu\nu} = \eta_{\mu\nu} + h_{\mu\nu} \tag{1.4}$$

where $\eta_{\mu\nu}$ is the usual flat Minkowski metric and $h_{\mu\nu}$ is a small correction. A gauge can always be chosen in which the n-dimensional field equations take the form

$$-\frac{1}{2}\eta^{\mu\nu}\partial_\mu\partial_\nu\bar{h}_{\sigma\tau} + O(h^2) = 8\pi G T_{\sigma\tau}$$

$$\eta^{\mu\nu}\partial_\mu\bar{h}_{\nu\sigma} = 0, \tag{1.5}$$

where

$$\bar{h}_{\sigma\tau} = h_{\sigma\tau} - \frac{1}{2}\eta_{\sigma\tau}\eta^{\mu\nu}h_{\mu\nu}, \quad \text{i.e.,}$$

$$h_{\sigma\tau} = \bar{h}_{\sigma\tau} - \frac{1}{n-2}\eta_{\sigma\tau}\eta^{\mu\nu}\bar{h}_{\mu\nu}. \tag{1.6}$$

The Newtonian limit is obtained by setting $T_{00} \approx \rho$, where ρ is the mass density; neglecting all other components of the stress–energy tensor; and ignoring time derivatives, which are suppressed by powers of v/c. The only nonzero component of $\bar{h}_{\mu\nu}$ is then

$$\bar{h}_{00} = -4\Phi, \tag{1.7}$$

where Φ is the Newtonian potential,

$$\nabla^2 \Phi = 4\pi G \rho. \tag{1.8}$$

In this limit, the geodesic equation

$$\frac{d^2 x^\rho}{ds^2} + \Gamma^\rho_{\mu\nu} \frac{dx^\mu}{ds} \frac{dx^\nu}{ds} = 0 \tag{1.9}$$

reduces to

$$\frac{d^2 x^i}{dt^2} - \frac{1}{2} \partial_i h_{00} = 0. \tag{1.10}$$

Combining (1.6) and (1.7), we see that

$$\frac{d^2 x^i}{dt^2} + \frac{2(n-3)}{n-2} \partial_i \Phi = 0. \tag{1.11}$$

In four dimensions, equation (1.11) gives the standard Newtonian equations of motion, and for $n > 4$ the standard equations may be obtained by rescaling the coupling constant G. In three spacetime dimensions, however, test particles experience no Newtonian force.

This absence of a Newtonian limit does not make the theory trivial: moving particles, for example, can still exhibit nontrivial scattering. In fact, point particle solutions in 2+1 dimensions are good models for parallel cosmic strings in 3+1 dimensions [134]. Cosmic strings are topological solitons that occur in certain gauge theories; it is conjectured that they may have formed during phase transitions in the early universe, where they could have played an important role in the formation of large-scale structure. A straight cosmic string along, say, the z axis is characterized by a stress tensor of the form $T_{00} = -T_{33} = \rho \delta(z)$, and the large tension in the z direction alters the Newtonian limit of ordinary (3+1)-dimensional general relativity. Indeed, the 'effective Newtonian mass density' for an object with pressures $T_{ii} = p_i$ is

$$\rho + \sum_i p_i, \tag{1.12}$$

which vanishes for a cosmic string. The dynamics of a set of such parallel strings may be described in terms of their behavior on the $z = 0$ plane, and for this purpose, (2+1)-dimensional gravity provides a useful model. This is a classical problem, however – at scales at which quantum gravity becomes important, a cosmic string can no longer be represented as a point defect – and I will have little to say about it in the remainder of this book.

1.2 Generalizations

There are several generalizations of (2+1)-dimensional general relativity
that restore local degrees of freedom, making the dynamics more like that
of realistic (3+1)-dimensional gravity. The quantization of these models is
not yet well understood, and they will not be a major topic of this book,
but they they warrant a brief introduction.

The first generalization is (2+1)-dimensional dilaton gravity, that is,
general relativity coupled to a scalar field φ (the dilaton). In its most
general form, the action can be written as [273]

$$I_{DG} = \int_M d^3x \sqrt{-g} \left(C[\varphi]R - \frac{\omega[\varphi]}{\varphi} g^{\mu\nu} \partial_\mu\varphi\partial_\nu\varphi + 2\varphi V[\varphi] \right),$$

$$(1.13)$$

where C, ω, and V are arbitrary functions of φ. Models of this kind arise
naturally in string theory,[*] with

$$C[\varphi] = \varphi, \quad \omega[\varphi] = -1, \quad V[\varphi] = \Lambda/2, \qquad (1.14)$$

while the choice

$$C[\varphi] = \varphi, \quad \omega[\varphi] = \omega_0, \quad V[\varphi] = 0 \qquad (1.15)$$

corresponds to three-dimensional Brans–Dicke–Jordan theory. In such
models, the scalar field φ becomes a local dynamical degree of freedom,
and a judicious choice of couplings can lead to a limit not unlike Newtonian gravity [30]. Many versions of dilaton gravity are known to have
black hole solutions (see, for example, [74, 75, 233]), but the quantization
of these models has not been studied in any great detail.

A second generalization is unique to 2+1 dimensions, where a 'gravitational Chern–Simons term'

$$I_{GCS} = -\frac{1}{32\pi G\mu} \int_M d^3x \, \epsilon^{\lambda\mu\nu} \Gamma^\rho_{\lambda\sigma} \left(\partial_\mu\Gamma^\sigma_{\rho\nu} + \frac{2}{3}\Gamma^\sigma_{\mu\tau}\Gamma^\tau_{\nu\rho} \right)$$

$$(1.16)$$

can be added to the gravitational action [91, 92]. This rather unusual-
looking term appears as a counterterm in the renormalization of quantum
field theory in a (2+1)-dimensional gravitational background [265, 262,
132]. The expression (1.16) does not appear to be generally covariant,
but it is, at least when the manifold M is closed: it may be checked that
an infinitesimal coordinate change merely adds a total derivative to the
Lagrangian, leaving the action unchanged.

Variation of the total action $I + I_{GCS}$ yields the equations of motion

$$G^{\mu\nu} + \mu^{-1}C^{\mu\nu} = 0, \qquad (1.17)$$

---

[*] In string theory, the field φ is usually denoted as $e^{-2\phi}$.

where $C^{\mu\nu}$ is the conformally invariant Cotton tensor,

$$C^{\mu\nu} = \frac{1}{\sqrt{g}}\epsilon^{\mu\rho\sigma}\nabla_\rho(R^\nu_\sigma - \frac{1}{4}\delta^\nu_\sigma R). \qquad (1.18)$$

The simple counting argument that gave us the number of degrees of freedom in Einstein gravity no longer holds: for such third-order equations of motion, the spatial metric and its time derivative must both be treated as configuration space variables with associated canonical momenta, and the analysis becomes more elaborate. Instead, as Deser, Jackiw, and Templeton first observed [91], the linearized equations of motion are those of a massive scalar field,

$$(\Box + \mu^2)\phi = 0, \qquad (1.19)$$

where

$$\phi = (\delta_{ij} + \hat{\partial}_i\hat{\partial}_j)h^{ij}, \qquad \text{with} \quad \hat{\partial}_i = \partial_i(-\nabla^2)^{-1/2}. \qquad (1.20)$$

The existence of such a massive excitation can be confirmed by looking at the effective interaction of static external sources: one finds a Yukawa attraction with an interaction energy

$$E = -\int d^2x \, T_{00}(-\nabla^2 + \mu^2)^{-1}T_{00}, \qquad (1.21)$$

as expected for a massive scalar 'graviton'. This model is commonly called topologically massive gravity ('topological', somewhat misleadingly, because the Chern–Simons term (1.16) is important in topology). Topologically massive gravity has been shown to be perturbatively renormalizable [94, 168], and a number of interesting classical solutions are known. Fairly little is known about the quantization of this system, however, although some progress has been made in understanding the canonical structure and the asymptotic states [46, 93, 137].

1.3 A note on units

It is customary in quantum gravity to express masses in terms of the Planck mass and lengths in terms of the the Planck length. In 2+1 dimensions the gravitational constant G has units of an inverse momentum, and the Planck mass (in units with $c = 1$) is

$$M_{Pl} = \frac{1}{G}, \qquad (1.22)$$

while the Planck length is

$$L_{Pl} = \hbar G. \qquad (1.23)$$

If a cosmological constant is present, $|\Lambda|^{-1/2}$ has units of length. The theory then has a dimensionless length scale,

$$\ell = \frac{1}{16\pi\hbar G|\Lambda|^{1/2}}. \tag{1.24}$$

Roughly speaking, this scale measures the radius of curvature of the universe.

Throughout this book, I will use units such that $16\pi G = 1$ and $\hbar = 1$, unless otherwise stated. This choice simplifies a number of equations, particularly those involving canonical momenta. In concrete applications, of course – if we are interested in the thermodynamic characteristics of black holes, for instance – it is important to restore factors of G and $\hbar$.

2
Classical general relativity in 2+1 dimensions

If we wish to quantize (2+1)-dimensional general relativity, it is important to first understand the classical solutions of the Einstein field equations. Indeed, many of the best-understood approaches to quantization start with particular representations of the space of solutions. The next three chapters of this book will therefore focus on classical aspects of (2+1)-dimensional gravity. Our goal is not to study the detailed characteristics of particular solutions, but rather to develop an understanding of the generic properties of the space of solutions.

In this chapter, I will introduce two fundamental approaches to classical general relativity in 2+1 dimensions. The first of these, based on the Arnowitt–Deser–Misner (ADM) decomposition of the metric, is familiar from (3+1)-dimensional gravity [9]; the main new feature is that for certain topologies, we will be able to find the general solution of the constraints. The second approach, which starts from the first-order form of the field equations, is also similar to a (3+1)-dimensional formalism, but the first-order field equations become substantially simpler in 2+1 dimensions.

In both cases, the goal is to set up the field equations in a manner that permits a complete characterization of the classical solutions. The next chapters will describe the resulting spaces of solutions in more detail. I will also derive the algebra of constraints in each formalism – a vital ingredient for quantization – and I will discuss the (2+1)-dimensional analogs of total mass and angular momentum.

2.1 The topological setting

Before plunging into a detailed analysis of the field equations, it is useful to ask a preliminary question: what spacetime topologies can occur in (2+1)-dimensional gravity?

9

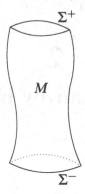

Fig. 2.1.   The manifold M has an initial boundary Σ^- and a final boundary Σ^+.

As we shall see below, the interesting cosmological solutions have nontrivial topologies, and to understand their structure, we shall need a number of mathematical tools. For readers unfamiliar with the fundamentals of the topology of manifolds, appendices A and B provide a brief summary of some relevant mathematics. Readers familiar with topology at the level of reference [204] should be able to skip these appendices, although they may serve as useful references for some particular applications.

It is helpful to divide our question into two parts. First, which three-manifolds admit Lorentzian metrics, that is, metrics that have the signature $(-++)$? Second, which of these manifolds admit solutions of the empty space Einstein field equations? Note that in 2+1 dimensions, this second question is more tractable than it might appear. As we saw in chapter 1, the vacuum field equations (with $\Lambda = 0$) require the metric to be flat, so we are really asking which three-manifolds admit flat Lorentzian metrics.

The first of these questions can be answered in full. In appendix B, it is shown that any noncompact three-manifold admits a Lorentzian metric, as does any closed three-manifold. ('Closed' means 'compact and without boundary'.) For compact manifolds with boundary, the problem becomes more interesting. Given a manifold with several boundary components, one can look for a Lorentzian metric for which these components are the past and future spatial boundaries of the universe, as in figure 2.1. Sorkin has shown that a three-manifold M admits a time-orientable Lorentzian metric with spacelike past boundary Σ^- and spacelike future boundary Σ^+ (and no other boundary components) if and only if

$$\chi(\Sigma^-) = \chi(\Sigma^+), \tag{2.1}$$

where $\chi(\Sigma)$ is the Euler number (or Euler–Poincaré characteristic) of Σ [242].

If Σ^- and Σ^+ are both connected, this result prohibits topology change, since the Euler number of a connected surface completely determines its topology. (Recall that the Euler number of an orientable genus g surface is $2 - 2g$.) Σ^- and Σ^+ need not be connected, however, so topology change is not completely ruled out. Instead, we obtain an interesting set of selection rules – for instance, a genus g surface can evolve into two surfaces of genus g_1 and g_2 only if $2 - 2g = (2 - 2g_1) + (2 - 2g_2)$, i.e., $g_1 + g_2 = g + 1$. We shall discuss quantum mechanical topology change in chapter 9; for now, let me simply note that a similar selection rule occurs in the path integral formalism.

We now turn to the more difficult problem, the question of which three-manifolds admit *flat* Lorentzian metrics. A complete answer to this question is not known, but a number of useful results can be found in the mathematics literature. For example, closed three-manifolds with flat Lorentzian metrics are understood fairly well [18, 113, 131]. All such manifolds are geodesically complete, making them interesting candidates for singularity-free spacetimes, and their fundamental groups can be described explicitly. Unfortunately, though, closed Lorentzian manifolds always contain closed timelike curves, and thus have limited value as models in classical general relativity.

Less is known about noncompact three-manifolds with flat Lorentzian metrics. A number of interesting examples are given in [103] and [104]; in particular, it is shown that any handlebody (that is, any 'solid genus g surface') can be given a geodesically complete flat Lorentzian metric. The resulting geometries are fairly bizarre – for instance, it is unlikely that they allow any time-slicing – and they could potentially serve as counterexamples for a number of plausible claims about (2+1)-dimensional gravity. These spacetimes have not yet been studied by physicists in any detail.

For our purposes, the most important result is a theorem due to Mess [195]. Suppose that M is a compact three-dimensional manifold with a flat, time-orientable Lorentzian metric and a purely spacelike boundary. Then M necessarily has the topology

$$M \approx [0, 1] \times \Sigma, \tag{2.2}$$

where Σ is a closed surface homeomorphic to one of the boundary components of M. This means that for spatially closed three-dimensional universes, topology change is forbidden by the field equations – the topology of spacetime is completely fixed by that of an initial spacelike slice. This powerful result greatly simplifies the study of classical (2+1)-dimensional cosmology, allowing us to ignore many of the more complicated spacetime topologies. Moreover, we shall see below that if Σ is any surface other than the two-sphere, a manifold with the topology (2.2) actually admits a large family of flat Lorentzian metrics, which can be described in considerable detail.

2.2 The ADM decomposition

The unified treatment of space and time is a cornerstone of general relativity. As a practical matter, however, it is sometimes useful to reintroduce an explicit – although largely arbitrary – division of spacetime into spatial and temporal directions. Such a division is described by the Arnowitt–Deser–Misner (ADM) formalism [9, 202].

The ADM decomposition of spacetime into space and time furnishes a natural setting for the initial value problem, and it underlies Wheeler's 'geometrodynamical' picture of classical general relativity as the dynamics of evolving spatial geometries. By providing a canonical description of the gravitational phase space, it leads to a Hamiltonian version of classical gravity, and suggests a useful approach to canonical quantization. In addition, the ADM approach simplifies the determination of conserved quantities in a spatially open universe, the gravitational analogs of total momentum and angular momentum. As we shall see in the next chapter, the global geometry of (2+1)-dimensional gravity leads to conservation laws that differ substantially from those in 3+1 dimensions.

We begin with a spacetime manifold with the topology $[0, 1] \times \Sigma$, where Σ is an open or closed two-surface. Such a spacetime represents a segment of a universe between an initial surface $\{0\} \times \Sigma$, which we assume to be spacelike, and a final surface $\{1\} \times \Sigma$, which we also assume to be spacelike. The ADM approach to (2+1)-dimensional general relativity starts with a slicing of the spacetime manifold M into constant-time surfaces Σ_t, each provided with a coordinate system $\{x^i\}$ and an induced metric $g_{ij}(t, x^i)$. To obtain the full three-geometry, we must describe the way nearby time slices Σ_t and Σ_{t+dt} fit together. To do so, we start at a point on Σ_t with coordinates x^i, and displace it infinitesimally in the direction normal to Σ_t. The resulting change in proper time can be written as

$$d\tau = N dt, \qquad (2.3)$$

where $N(t, x^i)$ is called the lapse function. This is not quite the whole story, however: in a generic coordinate system, such a displacement will not only shift the time coordinate, but will alter the spatial coordinates as well. To allow for this possibility, we write

$$x^i(t + dt) = x^i(t) - N^i dt, \qquad (2.4)$$

where $N^i(t, x^i)$ is called the shift vector. By the Lorentzian version of the Pythagoras theorem (see figure 2.2), the interval between points (t, x^i) and $(t + dt, x^i + dx^i)$ is then

$$ds^2 = -N^2 dt^2 + g_{ij}(dx^i + N^i dt)(dx^j + N^i dt). \qquad (2.5)$$

Equation (2.5) is the ADM form of the metric.

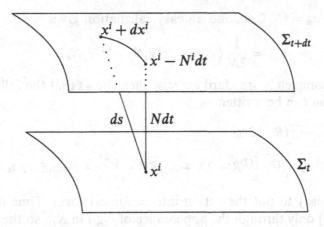

Fig. 2.2. The ADM decomposition is based on the Lorentzian version of the Pythagoras theorem.

It is customary in the ADM formalism to establish a new set of conventions that emphasize the role of the surface Σ. For the remainder of this section, spatial indices i, j, ... will be lowered and raised with the spatial metric g_{ij} and its inverse g^{ij}, and not with the full spacetime metric. Note that the components of g^{ij} are not simply the spatial components of the full three-metric $g^{\mu\nu}$: the inverse of (2.5) is

$$g^{\mu\nu} = \begin{pmatrix} -\dfrac{1}{N^2} & \dfrac{N^i}{N^2} \\ \dfrac{N^j}{N^2} & \left(g^{ij} - \dfrac{N^iN^j}{N^2}\right) \end{pmatrix}. \tag{2.6}$$

This convention can cause confusion at first, but it simplifies later notation.

The geometry of the slice Σ_t comprises two elements: the intrinsic geometry of the slice as a two-manifold, and the extrinsic geometry, which describes the embedding of Σ_t in the spacetime M. Just as the intrinsic geometry is determined by the behavior of vectors tangent to Σ_t under parallel transport, the extrinsic geometry is determined by the behavior of vectors normal to Σ_t. In particular, the extrinsic curvature K_{ij} of a surface Σ is defined by[*]

$$K_{\mu\nu} = -\nabla_\mu n_\nu + n_\mu n^\rho \nabla_\rho n_\nu \tag{2.7}$$

where ∇ is the full three-dimensional covariant derivative and n^μ is the unit normal to Σ. In the ADM decomposition (2.5), the normal to Σ_t has

---

[*] The reader should be warned that sign conventions for $K_{\mu\nu}$ vary, and that signs in a number of expressions in this chapter also depend on sign conventions for $g_{\mu\nu}$.

components $n_\mu = (N, 0, 0)$, and an easy calculation gives

$$K_{ij} = \frac{1}{2N}\left(\partial_t g_{ij} - {}^{(2)}\nabla_i N_j - {}^{(2)}\nabla_j N_i\right). \tag{2.8}$$

A long but completely standard exercise then shows that the full Einstein–Hilbert action can be written as

$$I = \int d^3x \sqrt{-g}\,(R - 2\Lambda)$$
$$= \int dt \int_\Sigma d^2x\, N\sqrt{{}^{(2)}g}\left[{}^{(2)}R - 2\Lambda + K_{ij}K^{ij} - K^2\right] + boundary\ terms. \tag{2.9}$$

It is now easy to put the action into canonical form. Time derivatives occur in (2.9) only through the appearance of $\partial_t g_{ij}$ in K_{ij}, so the canonical momenta are

$$\pi^{ij} = \frac{\partial \mathscr{L}}{\partial(\partial_t g_{ij})} = \sqrt{{}^{(2)}g}(K^{ij} - g^{ij}K). \tag{2.10}$$

Note that no time derivatives of N or N^i occur in the action; these variables have no canonical conjugates, and they will appear as Lagrange multipliers of constraints. Equation (2.10) can be inverted to give

$$K^{ij} = \frac{1}{\sqrt{{}^{(2)}g}}(\pi^{ij} - g^{ij}\pi), \tag{2.11}$$

and substituting back into (2.9), we can write the action as

$$I = \int dt \int_\Sigma d^2x \left(\pi^{ij}\partial_t g_{ij} - N\mathscr{H} - N_i\mathscr{H}^i\right), \tag{2.12}$$

where

$$\mathscr{H} = \frac{1}{\sqrt{{}^{(2)}g}}(\pi_{ij}\pi^{ij} - \pi^2) - \sqrt{{}^{(2)}g}\,({}^{(2)}R - 2\Lambda) \tag{2.13}$$

is known as the Hamiltonian constraint and

$$\mathscr{H}^i = -2{}^{(2)}\nabla_j \pi^{ij} \tag{2.14}$$

are the momentum constraints. Apart from numerical coefficients that depend on the dimension of spacetime, $\mathscr{H}$ and $\mathscr{H}^i$ have the same form as their (3+1)-dimensional counterparts. Note that the lapse function and the shift vector appear in the action only as Lagrange multipliers; their variation leads to the field equations $\mathscr{H} = 0$ and $\mathscr{H}^i = 0$.

Equation (2.12) is the standard Hamiltonian form of the gravitational action. Note that the 'Hamiltonian',

$$H = \int_\Sigma d^2x \left(N\mathscr{H} + N_i\mathscr{H}^i\right), \tag{2.15}$$

is proportional to the constraints, and thus vanishes on shell, that is, when the equations of motion are satisfied. This is the source of the common statement that the total energy of a closed universe is zero. We shall see in section 4 that for spatially open topologies, H must be supplemented by boundary terms that need not vanish on shell, permitting a natural definition of the total mass ('ADM mass') of a spacetime.

The Poisson brackets for (2+1)-dimensional gravity can be read from the action (2.12):

$$\{g_{ij}(x), \pi^{kl}(x')\} = \frac{1}{2}(\delta_i^k \delta_j^l + \delta_i^l \delta_j^k)\tilde{\delta}^2(x - x'), \tag{2.16}$$

where $\tilde{\delta}^2(x - x')$ is the metric-independent ('densitized') delta function, that is,

$$\int d^2x \, \tilde{\delta}^2(x - x')f(x') = f(x) \tag{2.17}$$

for any scalar function $f(x)$. For a finite-dimensional phase space, it is well known that the Poisson brackets define a symplectic structure, determined by a closed two-form Ω that can be written in local coordinates as

$$\Omega = \sum_i dp_i \wedge dq^i. \tag{2.18}$$

The infinite-dimensional generalization of (2.18) is

$$\Omega = \int_\Sigma d^2x \, \delta\pi^{ij} \wedge \delta g_{ij}, \tag{2.19}$$

where the variations δg_{ij} and $\delta\pi^{ij}$ are the infinite-dimensional analogs of the exterior derivatives in (2.18). This symplectic structure is critical for the quantum theory – among other things, it determines the canonical commutation relations – and we shall return to equation (2.19) and related expressions frequently throughout this book.

2.3 Reduced phase space and moduli space

It is well known that the metric and momentum variables of equation (2.12) are redundant: they describe both physical excitations and unobservable 'pure gauge' degrees of freedom that merely represent coordinate changes. The reduction to the space of true physical degrees of freedom requires two steps. First, we must solve the constraints $\mathcal{H} = 0$ and $\mathcal{H}^i = 0$. The positions and momenta that satisfy these constraints will lie on a submanifold of the phase space, the constraint surface. These solutions are still subject to gauge transformations, however, and as a second step we must factor out (or gauge-fix) these transformations. The resulting 'reduced phase space' is the space of true degrees of freedom of the theory [100, 101, 102, 135, 149].

In 3+1 dimensions, we do not know how to carry out such a program: the constraints are too difficult to solve. In 2+1 dimensions, however, a further decomposition of the phase space variables noticeably simplifies the ADM action. For closed 'cosmological' spacetimes, this decomposition, coupled with a clever choice of time-slicing, will allow us to reduce the constraints to a single differential equation, and will permit a fairly detailed description of the reduced phase space. As we shall see, this phase space is closely related to the Riemann moduli space $\mathcal{N}(\Sigma)$ of the surface Σ, a space that has been studied extensively by mathematicians.

We begin with a theorem from Riemann surface theory [110], which states that any metric on a compact surface Σ is conformal to a metric of constant (intrinsic) curvature k, where $k = 1$ for the two-sphere, $k = 0$ for the torus, and $k = -1$ for any surface of genus $g \geq 2$. This result, which is a version of the uniformization theorem discussed in appendix A, allows us to write the spatial metric g_{ij} as

$$g_{ij} = e^{2\lambda}\hat{g}_{ij}, \tag{2.20}$$

where $\hat{g}_{ij}$ is a constant curvature metric, unique up to diffeomorphisms of Σ. Moreover, the space of such constant curvature metrics modulo diffeomorphisms is known to be finite dimensional. This means that we may express any metric g_{ij} in the form

$$g_{ij} = e^{2\lambda}f^{*}\tilde{g}_{ij}, \tag{2.21}$$

where f is a diffeomorphism of Σ and $\tilde{g}_{ij}$ is one of a fixed finite-dimensional family of constant curvature metrics. If Σ is open, suitable boundary conditions are needed for this theorem to hold, and the implications for (2+1)-dimensional gravity are not yet fully understood; section 6 of this chapter includes a partial analysis.

The space of constant curvature metrics modulo diffeomorphisms is known as the moduli space $\mathcal{N}$ of Σ. It has dimension $6g - 6$ if the genus of Σ is $g > 1$, two if Σ is a torus ($g = 1$), and zero if Σ is a sphere ($g = 0$). This space of metrics will recur throughout this book, appearing, for example, in chapter 4 as the space of hyperbolic structures on Σ. It is roughly the same as Wheeler's superspace [285]; more precisely, it is 'conformal superspace', the space of metrics modulo diffeomorphisms and Weyl transformations.

Let us denote coordinates on $\mathcal{N}$ by m_{α}, and write $\tilde{g}_{ij} = \tilde{g}_{ij}(m_{\alpha})$. The variables conjugate to the m_{α} are, roughly speaking, parameters that label the traceless part of the momentum π^{ij}. More precisely, let $\tilde{\nabla}_i$ denote the covariant derivative compatible with the metric $\tilde{g}_{ij}$, set $\pi = g_{ij}\pi^{ij}$, and define

$$(PY)_{ij} = \tilde{\nabla}_i Y_j + \tilde{\nabla}_j Y_i - \tilde{g}_{ij}\tilde{g}^{kl}\tilde{\nabla}_k Y_l. \tag{2.22}$$

We can then decompose π^{ij} as

$$\pi^{ij} = e^{-2\lambda}\left(p^{ij} + \frac{1}{2}\tilde{g}^{ij}\pi + \sqrt{\tilde{g}}\tilde{g}^{ik}\tilde{g}^{jl}(PY)_{kl}\right), \tag{2.23}$$

where p^{ij} is a transverse traceless tensor density with respect to $\tilde{g}_{ij}$, that is,

$$\tilde{\nabla}_i p^{ij} = 0, \qquad \tilde{g}_{ij}p^{ij} = 0. \tag{2.24}$$

For a surface of genus $g > 1$, this decomposition is unique; for $g = 0$ or $g = 1$, the vector Y^i is not completely determined, but is unique up to the addition of a conformal Killing vector [202].

To obtain the symplectic structure in terms of these new variables, let us decompose an arbitrary variation δg_{ij} as

$$\delta g_{ij} = \nabla_i \delta\xi_j + \nabla_j \delta\xi_i + 2e^{2\lambda}\tilde{g}_{ij}\delta\lambda + e^{2\lambda}\delta\tilde{g}_{ij}, \tag{2.25}$$

where $\delta\xi^i$ is a vector field that generates an infinitesimal diffeomorphism. It is then easy to check that the symplectic form (2.19) becomes

$$\Omega = \int_\Sigma d^2x \left\{ \delta p^{ij} \wedge \delta\tilde{g}_{ij} + 2\delta\pi \wedge \delta\lambda \right.$$
$$\left. - 2\tilde{g}^{ij}\left[e^{-2\lambda}\sqrt{\tilde{g}}\left(\tilde{\Delta} + \frac{k}{2}\right)\delta Y_i + \frac{1}{2}\tilde{\nabla}_i\left(e^{-2\lambda}\delta\pi\right)\right] \wedge \delta\xi_j \right\}. \tag{2.26}$$

Thus π is conjugate to λ, p^{ij} is conjugate to $\tilde{g}_{ij}$, and Y_i is roughly conjugate to the diffeomorphism degrees of freedom of the spatial metric. We shall see below that the constraints determine Y_i and λ, and that corresponding gauge conditions fix their canonical conjugates, leaving a reduced phase space parametrized by $\tilde{g}_{ij}$ and p^{ij}.

In terms of these new variables, the momentum constraints $\mathcal{H}_i = 0$ become

$$\sqrt{\tilde{g}}\left(\tilde{\Delta} + \frac{k}{2}\right)Y_i + \frac{1}{2}e^{2\lambda}\tilde{\nabla}_i\left(e^{-2\lambda}\pi\right) = 0, \tag{2.27}$$

while the Hamiltonian constraint $\mathcal{H} = 0$ takes the form

$$0 = -\frac{1}{2}\frac{1}{\sqrt{\tilde{g}}}e^{-2\lambda}\pi^2 + 2\sqrt{\tilde{g}}\left[\tilde{\Delta}\lambda - \frac{k}{2}\right] + 2\sqrt{\tilde{g}}\Lambda e^{2\lambda} \tag{2.28}$$

$$+ \frac{1}{\sqrt{\tilde{g}}}e^{-2\lambda}\tilde{g}_{ik}\tilde{g}_{jl}\left(p^{ij} + \sqrt{\tilde{g}}\tilde{g}^{im}\tilde{g}^{jn}(PY)_{mn}\right)\left(p^{kl} + \sqrt{\tilde{g}}\tilde{g}^{kp}\tilde{g}^{lq}(PY)_{pq}\right),$$

where I have used the conformal transformation properties of the spatial curvature,

$$^{(2)}R[e^{2\lambda}\tilde{g}] = e^{-2\lambda}\left(^{(2)}R[\tilde{g}] - 2\tilde{\Delta}\lambda\right). \tag{2.29}$$

These expressions look more complicated than the original form (2.13)–(2.14), but they are actually quite a bit easier to solve.

To make further progress, we must fix a coordinate system; in particular, we must specify the splitting of spacetime into space and time. In 3+1 dimensions, it has been known for some time that a useful time-slicing is given by York's 'extrinsic time', in which spacetime is foliated by surfaces of constant mean extrinsic curvature $Tr\,K = g^{ij}K_{ij} = -T$ [297]. In (3+1)-dimensional general relativity, the question of the global existence of such a foliation is a difficult one, and some restrictions on the geometry and topology are necessary to ensure that the York time-slicing is possible. For spatially closed (2+1)-dimensional spacetimes with the topology $[0, 1] \times \Sigma$, on the other hand, Andersson, Moncrief, and Tromba have shown that such a slicing always exists, and that $Tr\,K$ can be used as a global time coordinate in a suitable region[†] of M [8].

In the ADM formalism, the extrinsic curvature for a classical solution of the Einstein equations is proportional to π^{ij}, and the York time is

$$T = g^{-1/2}g_{ij}\pi^{ij} = \tilde{g}^{-1/2}e^{-2\lambda}\pi. \tag{2.30}$$

With this choice of time-slicing, the momentum constraints (2.27) then imply that $Y_i = 0$. (In the genus 1 case, Y_i may be a conformal Killing vector, that is, $(PY)_{ij} = 0$, but by (2.23), we can take Y_i to vanish without loss of generality.) The remaining momentum degrees of freedom at constant T are thus given by the transverse traceless tensor p^{ij}, as might have been expected from the symplectic structure.

In Riemann surface theory, a symmetric rank two transverse traceless tensor is called a 'holomorphic quadratic differential': in local complex coordinates $ds^2 = e^{2\rho}dzd\bar{z}$, p^{ij} has one independent complex component, p_{zz}, and equation (2.24) requires this component to be holomorphic,

$$\partial_{\bar{z}}p_{zz} = 0. \tag{2.31}$$

It is a standard result of Riemann surface theory that the space of holomorphic quadratic differentials on a genus g surface is $(6g - 6)$-dimensional (two-dimensional if $g = 1$), and is in fact isomorphic to the cotangent space at $\tilde{g}_{ij}$ to the moduli space $\mathcal{N}$ [99, 120]. Up to spatial diffeomorphisms, the p^{ij} are therefore parametrized by $6g - 6$ coordinates

---

[†] The 'suitable region' here is the domain of dependence of a spacelike slice Σ, as described in appendix B.

p^α. These can be chosen so that the action (2.12) includes a standard kinetic term

$$I = \int dT \, p^\alpha \dot{m}_\alpha + \ldots \tag{2.32}$$

by setting

$$p^\alpha = \int_\Sigma d^2x \, p^{ij} \frac{\partial}{\partial m_\alpha} \tilde{g}_{ij}. \tag{2.33}$$

These p^α uniquely determine the momenta p^{ij}, and the pair of fields $\{p^{ij}, \tilde{g}_{ij}\}$ parametrize the cotangent bundle $T^*\mathcal{N}$ of moduli space. In fact, the p^α and m_α are conjugate variables: if we restrict the symplectic form (2.26) to the reduced phase space, it is easy to see that Ω simplifies to

$$\Omega = \sum_\alpha dp^\alpha \wedge dm_\alpha. \tag{2.34}$$

It remains for us to evaluate the Hamiltonian constraint (2.28) in the York time-slicing. Using the fact that Y^i now vanishes, we obtain an equation for the conformal factor λ,

$$\tilde{\Delta}\lambda - \frac{1}{4}(T^2 - 4\Lambda)e^{2\lambda}$$

$$+ \frac{1}{2}\left[\tilde{g}^{-1}\tilde{g}_{ij}(m_\alpha)\tilde{g}_{kl}(m_\alpha)p^{ik}(p^\alpha)p^{jl}(p^\alpha)\right]e^{-2\lambda} - \frac{k}{2} = 0. \tag{2.35}$$

This is an elliptic differential equation, and one can employ standard mathematical tools to investigate the existence of solutions. Moncrief has shown that when Σ is closed and $T^2 \geq 4\Lambda$, equation (2.35) has a unique solution, completely determining λ as a function on the reduced phase space parametrized by $\{m_\alpha, p^\alpha\}$ [202].

We can now insert the decomposition (2.21)–(2.23), the gauge choice (2.30), and our solutions of the constraints into the ADM action. The result is a reduced phase space action for the physical degrees of freedom,

$$I = \int dT \left\{ p^\alpha \frac{dm_\alpha}{dT} - H_{red}(m, p, T) \right\}, \tag{2.36}$$

where the reduced phase space Hamiltonian

$$H_{red}(m, p, T) = \int_{\Sigma_T} d^2x \, \sqrt{\tilde{g}} \, e^{2\lambda(m,p,T)} \tag{2.37}$$

is determined by solving equation (2.35) for λ. Geometrically, $H_{red}(T)$ is just the area of the surface $\text{Tr}K = -T$. For a complete solution, we also need the lapse and shift functions N and N^i, which can be obtained from the standard Einstein field equations once we have found g_{ij} and π^{ij}. Moncrief has shown that N and N^i are completely determined by the moduli m_α and the momenta p^α, as one would expect [202].

The dynamics of (2+1)-dimensional gravity in the York time-slicing thus reduces to that of a finite-dimensional system described by the action (2.36). In practice, this system is still too complicated for us to solve for most spacetime topologies. If Σ is a torus, however, we shall see in the next chapter that the general solution can be found.

2.4 Diffeomorphisms and conserved charges

The canonical decomposition provided by the ADM variables allows us to apply standard Hamiltonian techniques to (2+1)-dimensional gravity. In particular, the Poisson brackets (2.16) imply that

$$\{F[g,\pi], g_{ij}\} = -\frac{\delta F[g,\pi]}{\delta \pi^{ij}}$$

$$\{F[g,\pi], \pi^{ij}\} = \frac{\delta F[g,\pi]}{\delta g_{ij}} \tag{2.38}$$

for any functional F of the positions and momenta.[‡] The Hamiltonian form (2.12) of the action then leads to a variational principle

$$\dot{g}_{ij} = \frac{\delta H}{\delta \pi^{ij}} = -\{H, g_{ij}\}$$

$$\dot{\pi}^{ij} = -\frac{\delta H}{\delta g_{ij}} = -\{H, \pi^{ij}\} \tag{2.39}$$

with H as in equation (2.15), giving the standard form of Hamilton's equations.

As usual in a constrained system, the constraints $\mathcal{H} = \mathcal{H}^i = 0$ are not among the dynamical equations (2.39), but must be imposed separately. This should not be surprising: no time derivatives of N or N^i appear in the action, so we should expect the variational equations

$$\frac{\delta I}{\delta N} = \frac{\delta I}{\delta N^i} = 0 \tag{2.40}$$

to have an exceptional status. In fact, these equations are consistency conditions on initial data rather than dynamical equations. The existence of such constraints is a fundamental feature of general relativity, and for

---

[‡] Strictly speaking, this is true only up to possible boundary terms in the variation of F. This caveat is unimportant now, but will play a key role in the derivation of the conserved charges of (2+1)-dimensional gravity.

the remainder of this section we shall investigate their structure more carefully.

In gauge theories, constraints can typically be understood as generators of infinitesimal gauge transformations. Gravity is not quite a gauge theory – we shall see below that diffeomorphisms are not, strictly speaking, gauge transformations – but it is useful to develop the analog of this result. Let us start with the momentum constraints (2.14). Using the brackets (2.16), it is easy to verify that

$$\left\{ \int d^2x\, \xi_i(x)\mathcal{H}^i(x), g_{kl}(x') \right\} = -\left({}^{(2)}\nabla_k\xi_l + {}^{(2)}\nabla_l\xi_k \right)(x'),$$

$$(2.41)$$

which may be recognized as the equation for the variation of the metric under an infinitesimal spatial diffeomorphism $x^i \to x^i - \xi^i$. Similarly, a short calculation shows that

$$\left\{ \int d^2x\, \xi_i(x)\mathcal{H}^i(x), \pi^{kl}(x') \right\}$$

$$= \left(\xi^i\partial_i\pi^{kl} + \pi^{ik}\partial_i\xi^l + \pi^{il}\partial_i\xi^k + \pi^{kl}\partial_i\xi^i \right)(x'), \quad (2.42)$$

which is the correct transformation law for a tensor density under an infinitesimal diffeomorphism. The analogy with gauge theories is so far very close, and the momentum constraints can indeed be interpreted as generators of spatial diffeomorphisms.

We should expect more trouble with the Hamiltonian constraint (2.13). The remaining 'gauge' symmetry of general relativity consists of diffeomorphisms that move the surface Σ_t in time: $x^0 \to x^0 - \xi^0$, and consequently

$$\delta g_{ij} = -\left(\nabla_i\xi_j + \nabla_j\xi_i\right) = N_j\partial_i\xi^0 + N_i\partial_j\xi^0 + \xi^0\partial_t g_{ij}. \qquad (2.43)$$

(Here ∇ is the full three-dimensional covariant derivative.) Such transformations are in some sense dynamical, unlike ordinary gauge transformations that act on fields at a fixed time. This intertwining of dynamics and symmetry is an early sign of the 'problem of time' that will return to plague us when we try to quantize the theory.

It is not hard to check that

$$\left\{ \int d^2x\, \xi(x)\mathcal{H}(x), g_{kl}(x') \right\} = -\frac{2}{\sqrt{{}^{(2)}g}} \xi(x')(\pi_{kl} - g_{kl}\pi)(x').$$

$$(2.44)$$

On shell – when the dynamical equation (2.10) is satisfied – this becomes

$$\left\{ \int d^2x\, \xi(x)\mathcal{H}(x), g_{kl}(x') \right\} = -\frac{\xi}{N}(x')\left(\partial_t g_{kl} - {}^{(2)}\nabla_k N_l - {}^{(2)}\nabla_l N_k \right)(x').$$

$$(2.45)$$

Setting $\xi^0 = -\xi/N$, we can recognize this expression as a transformation of the form (2.43), up to a purely spatial diffeomorphism of the form (2.41) with $\xi_l = -\xi N_l/N$. The Hamiltonian constraint $\mathcal{H}$ thus generates diffeomorphisms in the time direction – deformations of the surface Σ_t – but only modulo the dynamical equations of motion. The same phenomenon can be shown to hold when one considers transformations of the momenta: the Hamiltonian constraint generates transformations that can be identified with diffeomorphisms, but only on shell.

Let us next examine the Poisson brackets of the constraints among themselves. The result is most easily expressed in terms of the generator

$$\mathcal{G}[\xi, \xi^i] = \int_\Sigma d^2x \left(\xi \mathcal{H} + \xi^i \mathcal{H}_i \right). \tag{2.46}$$

A long but routine calculation shows that up to possible boundary terms coming from integration by parts,

$$\{\mathcal{G}[\xi_1, \xi_1^i], \mathcal{G}[\xi_2, \xi_2^i]\} = \mathcal{G}[\xi_3, \xi_3^i], \tag{2.47}$$

with

$$\xi_3 = \xi_1^i \partial_i \xi_2 - \xi_2^i \partial_i \xi_1$$
$$\xi_3^k = \xi_1^i \partial_i \xi_2^k - \xi_2^i \partial_i \xi_1^k + g^{ki}(\xi_1 \partial_i \xi_2 - \xi_2 \partial_i \xi_1). \tag{2.48}$$

For purely spatial diffeomorphisms, $\xi_1 = \xi_2 = 0$, and equation (2.48) for ξ_3^k reduces to the ordinary Lie bracket of vector fields ξ_1^i and ξ_2^k. The algebra (2.47) may then be recognized as the Lie algebra of spatial diffeomorphisms. For transformations that are not purely spatial, on the other hand, our brackets do not even form a Lie algebra, since ξ_3^k explicitly involves the metric – we have 'structure functions' rather than structure constants. Nonetheless, Teitelboim has shown that the general structure of (2.47)–(2.48) gives a universal description of the algebra of surface deformations, depending solely on general covariance and not on the detailed dynamics of gravity; only the specific metric appearing in (2.48) varies from model to model [253].

When the time-slice Σ is an open surface, added complications occur in the ADM formalism, arising from the need to include boundary terms in the action. While these terms seem at first to be no more than a mathematical annoyance, their careful treatment can lead to important physical information. In particular, we shall now derive the conserved asymptotic 'charges' of (2+1)-dimensional gravity – the total mass and angular momentum – from the ADM boundary terms.

The need for boundary terms can be best understood by starting with the Hamiltonian form of the action and considering the variational principle

$$\frac{\delta I}{\delta g_{ij}} = \frac{\delta I}{\delta \pi^{ij}} = 0. \tag{2.49}$$

To derive the field equations, we normally assume that we can freely integrate by parts: the full variation of the action has the general form

$$\delta I = \int dt \int_\Sigma d^2x \left(A^{ij}\delta g_{ij} + B_{ij}\delta\pi^{ij} \right)$$
$$+ \int dt \int_{\partial\Sigma} d\sigma^i \left(A^j\delta g_{ij} + g_{ij}B_k\delta\pi^{jk} \right), \quad (2.50)$$

but we ordinarily simply ignore the boundary variations. This procedure may make sense in a Lagrangian formalism, where we can restrict our attention to variations δg_{ij} and $\delta\pi^{ij}$ that fall off fast enough at spatial infinity to kill the boundary terms. If we hope to derive Hamilton's equations from the canonical action principle, however, we can no longer be so cavalier. The functional derivatives appearing in the Poisson brackets (2.39) involve arbitrary fields, and there is no possibility of restricting the variations. If boundary terms are present, brackets involving H are simply not well-defined; in the terminology of Regge and Teitelboim, the Hamiltonian is not differentiable [229]. When this happens, the only known way to make the Hamiltonian formalism self-consistent is to add terms to H to cancel the boundary variations.

More generally, we can consider the generators of 'gauge' transformations introduced earlier,

$$\mathcal{G}[\xi, \xi^i] = \int_\Sigma d^2x \left(\xi\mathcal{H} + \xi^i\mathcal{H}_i \right), \quad (2.51)$$

of which the Hamiltonian H is a special case. For at least some choices of parameters ξ and ξ^i, we would like the Poisson brackets of $\mathcal{G}[\xi, \xi^i]$ to be well-defined. If ξ and ξ^i fall off rapidly enough at spatial infinity, of course, boundary terms will cause no problems. But by studying the behavior of the generators when ξ and ξ^i do not vanish at infinity, we can also explore the asymptotic symmetries of the theory.

Let us start by considering the variation of the term $\xi\mathcal{H}$ in $\mathcal{G}$. By equation (2.13), the only piece of $\mathcal{H}$ involving derivatives of the fields, and therefore requiring integration by parts, is the term $\sqrt{{}^{(2)}g}\,{}^{(2)}R$. Rather than explicitly working out the variation, we can use a trick suggested by Henneaux [147]. By the Gauss–Bonnet theorem, the integral

$$\frac{1}{4\pi} \int d^2x \sqrt{{}^{(2)}g}\,{}^{(2)}R$$

for a *closed* two-dimensional surface is a topological invariant, the Euler number, which cannot change under smooth variations of the metric. This suggests that we should be able to write an arbitrary variation in the form

$$\delta \left(\sqrt{{}^{(2)}g}\,{}^{(2)}R \right) = \partial_i\delta v^i, \quad (2.52)$$

since the integral of the right-hand side over a closed surface vanishes identically. This is in fact possible – v^i is a complicated function of the metric, and it is not globally single-valued (although its variation is), but it can be shown to exist. For an *open* surface, we therefore have

$$\delta \int_\Sigma d^2x \, \xi \sqrt{^{(2)}g} \, ^{(2)}R = \int_\Sigma d^2x \, \partial_i(\xi \delta v^i) + \ldots = \int_{\partial\Sigma} d\phi \, \xi(\delta v^\perp) + \ldots, \tag{2.53}$$

where the superscript $\perp$ denotes the direction orthogonal to the boundary $\partial\Sigma$ and I have kept only the boundary term.

Turning next to the term $\xi_i \mathscr{H}^i$ in $\mathscr{G}$, it is easy to show that the corresponding boundary term takes the form

$$\delta \int_\Sigma d^2x \, \xi_i \, ^{(2)}\nabla_j \pi^{ij} = \int_{\partial\Sigma} d\phi \left(\xi_i \delta \pi^{i\perp} + (\xi^l \pi^{k\perp} - \frac{1}{2}\xi^\perp \pi^{kl})\delta g_{kl} \right) + \ldots. \tag{2.54}$$

Combining (2.53) and (2.54), we find a total boundary variation

$$\delta \mathscr{G}[\xi, \xi^i] = \int_{\partial\Sigma} d\phi(-\xi \delta v^\perp - 2\xi^k \delta(g_{kl}\pi^{l\perp}) + \xi^\perp \pi^{kl} \delta g_{kl}) + \ldots, \tag{2.55}$$

which must be eliminated in order to have well-defined Poisson brackets involving $\mathscr{G}$.

Now suppose that ξ and ξ^i go to fixed values, say $\bar\xi$ and $\bar\xi^i$, at spatial infinity. The first two terms in (2.55) are then of the form $-\delta\bar{\mathscr{G}}$, where

$$\bar{\mathscr{G}}[\bar\xi, \bar\xi^i] = \int_{\partial\Sigma} d\phi \left(\bar\xi v^\perp + 2\bar\xi^k g_{kl}\pi^{l\perp} \right)$$

$$= \int_\Sigma d^2x \left(\bar\xi \sqrt{^{(2)}g} \, ^{(2)}R + 2\, ^{(2)}\nabla_i(\bar\xi_j \pi^{ij}) \right). \tag{2.56}$$

We can thus cancel most of the boundary variation by supplementing the 'volume' generator $\mathscr{G}$ with a new 'boundary' generator $\bar{\mathscr{G}}$.

The last term in equation (2.55) cannot be treated in this manner, however, and can only be eliminated by demanding that

$$\bar\xi^\perp = 0. \tag{2.57}$$

Recalling that the ξ^i parametrize spatial diffeomorphisms, this condition implies that not all asymptotic diffeomorphisms are admissible as symmetries of the theory. With the restriction (2.57), however, the quantity

$$\mathscr{G}'[\xi, \xi^i] = \mathscr{G}[\xi, \xi^i] + \bar{\mathscr{G}}[\bar\xi, \bar\xi^\parallel] \tag{2.58}$$

has well-defined Poisson brackets even for spatially open geometries, and can be taken as the complete generator of symmetries. In particular, $\mathscr{G}'$

will obey the algebra (2.47)–(2.48) of surface deformations, now with no ambiguities from possible boundary terms.

Geometrically, the restriction (2.57) means that we only allow diffeomorphisms that take an ideal circle at infinity to itself. This limitation is a reflection of the conical geometry of (2+1)-dimensional gravity. In 3+1 dimensions, the spacetime surrounding a collection of isolated sources is asymptotically flat, and has the full Poincaré group as its group of symmetries at spatial infinity. In 2+1 dimensions, on the other hand, we shall see in the next chapter that the corresponding solution is 'asymptotically conical'. Such a conical spacetime has a much smaller group of symmetries, consisting solely of spatial rotations, described by $\bar{\xi}^{\parallel}$, and time translations, described by $\bar{\xi}$ [87, 147]. For the case of a nonvanishing cosmological constant, the more complicated asymptotic symmetries are described in references [41, 42, 82].

So far, the boundary term $\mathscr{G}$ has played a limited role in our analysis, having been introduced primarily for mathematical consistency. Note, however, that while the 'volume' term $\mathscr{G}$ vanishes on shell, $\bar{\mathscr{G}}$ can be nonzero for classical solutions of the field equations. $\bar{\mathscr{G}}[\bar{\xi}, \bar{\xi}^{\parallel}]$ is therefore a candidate for a set of nontrivial global charges, analogous to the ordinary Hamiltonian and momentum in field theory.

Moreover, if ξ and ξ^i are asymptotically constant, equations (2.47)–(2.48) tell us that these charges commute on shell. In particular, if the lapse and shift functions go to constant values at spatial infinity, the $\bar{\mathscr{G}}[\bar{\xi}, \bar{\xi}^{\parallel}]$ will commute with the Hamiltonian (2.15), thus behaving as genuine constants of motion. In the next chapter, we shall see that these constants can be physically interpreted as the total mass and angular momentum of an isolated system.

2.5 The first-order formalism

The ADM approach is based on a natural choice of variables, the spatial metric and its canonically conjugate momentum. As we shall see in chapter 5, this choice leads to a particular approach to quantization. There is another equally natural choice of classical variables, however, that leads to a rather different quantum theory.

The first-order formalism takes as its fundamental variables a local frame ('triad' or 'dreibein') $e_\mu{}^a$, related to the metric by

$$\eta_{ab}e_\mu{}^a e_\nu{}^b = g_{\mu\nu}, \tag{2.59}$$

and a spin connection $\omega_\mu{}^{ab}$ (see appendix C). As in the Palatini formalism, e and ω are considered independent variables, to be varied separately in the action. In 2+1 dimensions, both e and ω can be treated as one-forms on the spacetime M:

$$e^a = e_\mu{}^a dx^\mu, \qquad \omega^a = \frac{1}{2}\epsilon^{abc}\omega_{\mu bc}dx^\mu. \tag{2.60}$$

The conventional Einstein–Hilbert action now takes the form

$$I = -2\int_M \left\{ e^a \wedge \left(d\omega_a + \frac{1}{2}\epsilon_{abc}\omega^b \wedge \omega^c \right) + \frac{\Lambda}{6}\epsilon_{abc}e^a \wedge e^b \wedge e^c \right\}. \tag{2.61}$$

The Euler–Lagrange equation coming from the variation of ω is

$$T_a = de_a + \epsilon_{abc}\omega^b \wedge e^c = 0, \tag{2.62}$$

the conventional 'no torsion' condition that ensures that the connection ω is compatible with the triad e. As long as the triad $e_\mu{}^a$ is invertible, equation (2.62) can be solved for ω in terms of e, giving the standard relation

$$\omega_\mu{}^a = \epsilon^{abc}e^\nu{}_c(\partial_\mu e_{\nu b} - \partial_\nu e_{\mu b}) - \frac{1}{2}\epsilon^{bcd}(e^\nu{}_b e^\rho{}_c \partial_\rho e_{\nu d})e_\mu{}^a. \tag{2.63}$$

This, in turn, implies that the full covariant derivative of e vanishes,

$$D_\mu e_\nu{}^a = \partial_\mu e_\nu{}^a - \Gamma^\rho_{\mu\nu}e_\rho{}^a + \epsilon^{abc}\omega_{\mu b}e_{\nu c} = 0, \tag{2.64}$$

where the $\Gamma^\rho_{\mu\nu}$ are the Christoffel symbols for the metric (2.59). The equation coming from the variation of e is then

$$R_a = d\omega_a + \frac{1}{2}\epsilon_{abc}\omega^b \wedge \omega^c = -\frac{\Lambda}{2}\epsilon_{abc}e^b \wedge e^c, \tag{2.65}$$

which is the condition that the spacetime have constant curvature. In 2+1 dimensions, we have seen that this is equivalent to the empty space Einstein field equations.[§]

Note that the equivalence of the first- and second-order formalisms is not quite complete: the first-order field equations yield a solution of the second-order equations only when the triad $e_\mu{}^a$ is invertible. For classical applications, the requirement of invertibility is a natural one, but noninvertible triads could in principle be important in the quantum theory.

Up to possible boundary terms, which will be discussed later, the action (2.61) is invariant under two sets of gauge symmetries: the local Lorentz

---

[§] A similar formalism exists in 3+1 dimensions. There, however, the first-order Lagrangian has a term of the form $\epsilon^{abcd}e_a \wedge e_b \wedge R_{cd}$, and the field equations determine $\epsilon^{abcd}e_b \wedge R_{cd}$, which is proportional to the Einstein tensor rather than the full curvature tensor.

transformations,

$$\delta e^a = \epsilon^{abc} e_b \tau_c$$
$$\delta \omega^a = d\tau^a + \epsilon^{abc} \omega_b \tau_c, \tag{2.66}$$

and the 'local translations',

$$\delta e^a = d\rho^a + \epsilon^{abc} \omega_b \rho_c$$
$$\delta \omega^a = -\Lambda \epsilon^{abc} e_b \rho_c. \tag{2.67}$$

The action is also manifestly invariant under spacetime diffeomorphisms, but this is not an independent symmetry. Indeed, using the identity

$$\mathscr{L}_\xi \sigma = d(\xi \cdot \sigma) + \xi \cdot d\sigma \tag{2.68}$$

for the Lie derivative $\mathscr{L}_\xi$ of any one-form σ, we easily find that

$$\mathscr{L}_\xi e^a = d(\xi \cdot e^a) + \epsilon^{abc} \omega_b (\xi \cdot e_c) + \epsilon^{abc} e_b (\xi \cdot \omega_c)$$
$$+ \textit{terms proportional to the equations of motion}$$
$$\mathscr{L}_\xi \omega^a = d(\xi \cdot \omega^a) + \epsilon^{abc} \omega_b (\xi \cdot \omega_c) - \Lambda \epsilon^{abc} e_b (\xi \cdot e_c)$$
$$+ \textit{terms proportional to the equations of motion}, \tag{2.69}$$

which may be recognized as the transformations (2.66) and (2.67) with (field-dependent) parameters

$$\rho^a = \xi \cdot e^a, \qquad \tau^a = \xi \cdot \omega^a. \tag{2.70}$$

This relationship is valid, once again, only when the triad is invertible. Observe also that the proof of equivalence holds only for 'small' diffeomorphisms, that is, diffeomorphisms that can be built up from infinitesimal transformations. The 'large' diffeomorphisms, those that cannot be smoothly deformed to the identity, must be treated separately; they will play an important role in the quantum theory.

This representation of diffeomorphisms as gauge transformations is not unique to (2+1)-dimensional general relativity, but is characteristic of so-called topological field theories. The possibility of exchanging the complicated diffeomorphism group for a much simpler group of pointwise gauge transformations is one of the principle reasons that the quantization of (2+1)-dimensional gravity is comparatively straightforward.

As Achúcarro and Townsend first pointed out, this gauge theory-like property of (2+1)-dimensional gravity can be explained by the observation that the first-order action (2.61) *is*, in fact, that of a gauge theory, specifically a Chern–Simons theory [1, 287]. Let $A = A_\mu{}^a T_a dx^\mu$ be the vector potential for a gauge theory with a gauge group G that has generators

T_a. (Mathematically, A is a connection one-form on a principle G bundle over M, as described in appendix C.) The Chern–Simons action for A is defined to be

$$I_{CS}[A] = \frac{k}{4\pi} \int_M Tr \left\{ A \wedge dA + \frac{2}{3} A \wedge A \wedge A \right\}, \qquad (2.71)$$

where k is a coupling constant and Tr is a nondegenerate, invariant bilinear form on the Lie algebra of G (for example, the matrix trace in a suitable representation). The Euler–Lagrange equations for this action are

$$F[A] = dA + A \wedge A = 0. \qquad (2.72)$$

In physical terms, the field strength of A vanishes; in mathematical terms, A is a flat connection. Note that this does *not* necessarily mean that A is trivial: if spacetime M is not simply connected, even a potential with vanishing field strength F can give rise to nontrivial Aharonov–Bohm phases around noncontractible loops.

Let us now take G to be the Poincaré group $ISO(2,1)$, with generators $\mathscr{J}^a$ of Lorentz transformations and $\mathscr{P}^a$ of translations, obeying the standard commutation relations

$$\left[\mathscr{J}^a, \mathscr{J}^b \right] = \epsilon^{abc} \mathscr{J}_c$$

$$\left[\mathscr{J}^a, \mathscr{P}^b \right] = \epsilon^{abc} \mathscr{P}_c$$

$$\left[\mathscr{P}^a, \mathscr{P}^b \right] = 0. \qquad (2.73)$$

The simplest invariant bilinear form on this algebra is given by

$$Tr(\mathscr{J}^a \mathscr{P}^b) = \eta^{ab}, \qquad Tr(\mathscr{J}^a \mathscr{J}^b) = Tr(\mathscr{P}^a \mathscr{P}^b) = 0. \qquad (2.74)$$

If we write our gauge potential as

$$A = e^a \mathscr{P}_a + \omega^a \mathscr{J}_a, \qquad (2.75)$$

then it is straightforward to verify that up to boundary terms, the Chern–Simons action (2.71) is precisely the first-order gravitational action (2.61) with $\Lambda = 0$ and

$$k = -\frac{1}{4G}, \qquad (2.76)$$

where for future convenience I have reinstated the coupling constant G. Moreover, it is not hard to check that the infinitesimal transformations (2.66)–(2.67) (with $\Lambda = 0$) are precisely the ordinary $ISO(2,1)$ gauge transformations of A.

A similar construction is possible when $\Lambda \neq 0$. For $\Lambda = -1/\ell^2 < 0$, the pair of one-forms

$$A^{(\pm)a} = \omega^a \pm \frac{1}{\ell}e^a \tag{2.77}$$

together constitute an $SO(2,1) \times SO(2,1)$ gauge potential, whose Chern–Simons action

$$I[A^{(+)}, A^{(-)}] = I_{CS}[A^{(+)}] - I_{CS}[A^{(-)}] \tag{2.78}$$

with

$$k = -\frac{\ell}{4G} \tag{2.79}$$

is again equivalent to (2.61). If $\Lambda > 0$, the complex one-form

$$A^a = \omega^a + i\sqrt{\Lambda}e^a \tag{2.80}$$

may be viewed as an $SL(2,\mathbf{C})$ gauge potential, whose Chern–Simons action is once again equivalent to the first-order gravitational action. For either sign of Λ, it is straightforward to show that the transformations (2.66)–(2.67) are again gauge transformations of the potential. Since the quantization of Chern–Simons theories is fairly well understood, this description gives us an obvious starting point for quantum gravity.

2.6 Boundary terms and the WZW action

Let us now return to the question of gauge invariance and boundary terms. Ordinarily, we construct a gauge-invariant action by choosing a Lagrangian that depends on the potential A only through the field strength F. The action (2.71) depends explicitly on A, and is not obviously gauge invariant. However, a straightforward computation shows that under the gauge transformation

$$A = g^{-1}dg + g^{-1}\tilde{A}g, \tag{2.81}$$

the Chern–Simons action transforms as

$$I_{CS}[A] = I_{CS}[\tilde{A}] - \frac{k}{4\pi} \int_{\partial M} Tr\left\{(dgg^{-1}) \wedge \tilde{A}\right\}$$
$$- \frac{k}{12\pi} \int_M Tr\left\{(g^{-1}dg) \wedge (g^{-1}dg) \wedge (g^{-1}dg)\right\}. \tag{2.82}$$

For closed manifolds, the boundary integral on the right-hand side of (2.82) does not appear, and the last term is a topological invariant, the winding number of the gauge transformation g [92, 165, 289]. For an appropriate choice of the coupling constant k, this winding number term is always an integral multiple of 2π, so the quantity $\exp\{iI_{CS}\}$ that occurs in a path integral is, indeed, gauge invariant. The restriction on k depends on the normalization of the bilinear form Tr. For most compact gauge groups with a standard choice of normalization, one must require that k be an integer. For gravity, at least in the case of Lorentzian signature metrics, there appears to be no quantization requirement, and the coupling constant G can be arbitrary.

If M is not closed, on the other hand, the gauge invariance of the Chern–Simons action is spoiled by the boundary term. Moreover, the expression (2.71) for the action is no longer quite correct, but must be supplemented by an appropriate boundary term, much as we had to modify the diffeomorphism generators in section 4. Indeed, under a variation of the potential A, the Chern–Simons action gives

$$\delta I_{CS}[A] = -\frac{k}{4\pi} \int_{\partial M} Tr A \wedge \delta A$$
$$+ \ terms \ proportional \ to \ the \ equations \ of \ motion, \quad (2.83)$$

and the boundary term must be eliminated if the action is to have genuine extrema.

A similar phenomenon occurs in the simpler case of scalar field theory, and can offer us guidance on how to proceed. Suppose we write the action for a scalar field ϕ in the form

$$I_M[\phi] = \frac{1}{2} \int_M d^n x \sqrt{-g} \ \phi \Delta \phi, \qquad (2.84)$$

where Δ is the scalar Laplacian. Let h denote the induced metric on ∂M, and let n^μ be the unit normal at the boundary. A variation of ϕ then yields

$$\delta I_M[\phi] = \int_M d^n x \sqrt{-g} \ \delta\phi \Delta\phi + \frac{1}{2} \int_{\partial M} d^{n-1} x \sqrt{h} \ (\phi n^\mu \nabla_\mu \delta\phi - \delta\phi \ n^\mu \nabla_\mu \phi), \qquad (2.85)$$

and the surface term does not vanish for either Dirichlet or Neumann boundary conditions. The action therefore has no genuine extrema, except in the trivial case that both ϕ and its normal derivative vanish at ∂M. The cure is obvious, however: if we choose boundary conditions in which ϕ is fixed at the boundary, we must add a surface term

$$I_{\partial M}[\phi] = -\frac{1}{2} \int_\Sigma d^{n-1} x \sqrt{h} \phi n^\mu \nabla_\mu \phi \qquad (2.86)$$

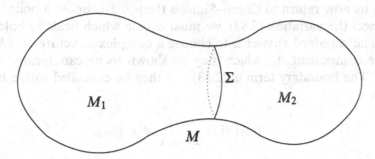

Fig. 2.3. A path integral 'sews' properly if the amplitude on the manifold M can be recovered from amplitudes for M_1 and M_2 by integrating over boundary data on Σ.

to obtain a total action

$$I'[\phi] = I_M[\phi] + I_{\partial M}[\phi] = -\frac{1}{2}\int_M d^n x\sqrt{-g}\,\nabla^\mu\phi\nabla_\mu\phi. \qquad (2.87)$$

It is easy to check that the variation of $I_{\partial M}$ exactly cancels the boundary term in (2.85) as long as $\delta\phi = 0$ at the boundary. Similarly, if we choose to fix the normal derivative of ϕ at ∂M, we should take our action to be $I''[\phi] = I_M[\phi] - I_{\partial M}[\phi]$. There is no choice of action that has extrema for both Neumann and Dirichlet boundary conditions; to select the appropriate action, we must first decide what boundary data we wish to hold fixed.

The need for such boundary terms is even more striking in the quantum theory, where it can be shown that they are necessary in order to ensure that partition functions 'sew' properly. Consider a manifold M split along some surface Σ into two components M_1 and M_2, as in figure 2.3. Letting ϕ_0 denote the value of the field ϕ on Σ, we can define the quantum partition function by the path integral

$$Z_{M_1}[\phi_0] = \int [d\phi]\, e^{iI_{M_1}[\phi]}, \qquad (2.88)$$

which will typically depend on the boundary value ϕ_0. The 'sewing' condition, the path integral version of summation over a complete set of intermediate states, is that

$$Z_M = \int [d\phi_0] Z_{M_1}[\phi_0] Z_{M_2}[\phi_0]. \qquad (2.89)$$

It may be shown by explicit computation that for both free field theories and Chern–Simons theories, this 'sewing' relation holds only when the appropriate boundary terms are included in the action [65, 291].

Let us now return to Chern–Simons theory. To choose a boundary term to cancel the variation (2.83), we must decide which fields to hold fixed at ∂M. The standard answer is to choose a complex structure on ∂M and to fix the component A_z, which may be shown to be canonically conjugate to $A_{\bar{z}}$. The boundary term in (2.83) can then be cancelled with a boundary action

$$I_{\partial M}[A] = \frac{k}{2\pi} \int_{\partial M} d^2x \, A_z A_{\bar{z}}. \tag{2.90}$$

It is now straightforward to show that under the gauge transformation (2.81), the modified Chern–Simons action $I'_{CS} = I_{CS} + I_{\partial M}$ transforms as

$$I'_{CS}[A] = I'_{CS}[\tilde{A}] + k I^+_{WZW}[g, \tilde{A}_z], \tag{2.91}$$

where $I^+_{WZW}[g, \tilde{A}_z]$ is the action of a chiral Wess–Zumino–Witten (or Wess–Zumino–Novikov–Witten) model on the boundary ∂M,

$$I^+_{WZW}[g, \tilde{A}_z] =$$
$$\frac{1}{4\pi} \int_{\partial M} Tr \left(g^{-1}\partial_z g \, g^{-1}\partial_{\bar{z}} g - 2g^{-1}\partial_{\bar{z}} g \tilde{A}_z \right) + \frac{1}{12\pi} \int_M Tr \left(g^{-1}dg \right)^3. \tag{2.92}$$

This result indicates that the number of physical degrees of freedom of a Chern–Simons theory – and in particular of (2+1)-dimensional gravity – depends strongly on whether spacetime has a boundary. For a closed universe, the solutions of a Chern–Simons theory are the flat connections modulo gauge transformations, and the space of such solutions is known to be finite dimensional. In the presence of a boundary, on the other hand, gauge invariance is broken at the boundary, and the 'would-be pure gauge' degrees of freedom g at ∂M become dynamical, adding an infinite-dimensional space of inequivalent solutions. In condensed matter physics these new degrees of freedom have been used to describe the fractional quantum Hall effect (see, for example, [282, 283, 284]), and we shall see in chapter 12 that the boundary degrees of freedom in quantum gravity provide a possible explanation for the thermodynamic properties of black holes.

While the boundary degrees of freedom described here are most easily derived in the Chern–Simons formalism, Balachandran and his collaborators have argued that they can be seen in the metric formulation of general relativity as well [20]. Consider, for example, the momentum constraints (2.14) on a spatial surface Σ with boundary $\partial\Sigma$. As in section 4 of this chapter, these constraints are *almost* the generators of diffeomorphisms, but the Poisson brackets (2.41) are now spoiled by the presence of the boundary. Indeed, starting with the expression (2.41) for the constraints and integrating by parts, we find that

$$\left\{ \int d^2x\, \xi_i(x)\mathscr{H}^i(x), g_{kl}(x') \right\} =$$

$$\left\{ \int d^2x\, ({}^{(2)}\nabla_i\xi_j + {}^{(2)}\nabla_j\xi_i)\pi^{ij}(x), g_{kl}(x') \right\} - 2\left\{ \int_{\partial\Sigma} d\phi\, \xi_i\pi^{i\perp}(x), g_{kl}(x') \right\}$$

$$= -\left({}^{(2)}\nabla_k\xi_l + {}^{(2)}\nabla_l\xi_k \right)(x') - 2\left\{ \int_{\partial\Sigma} d\phi\, \xi_i(x)\pi^{i\perp}(x), g_{kl}(x') \right\}.$$
(2.93)

Now, in ordinary canonical gravity the neglect of 'pure gauge' degrees of freedom is justified by the fact that the generator of diffeomorphisms is proportional to a constraint, and therefore vanishes on shell. In the absence of the boundary term in equation (2.93), we would be able to identify the metrics g_{kl} and $g'_{kl} = g_{kl} - \nabla_k\xi_l - \nabla_l\xi_k$, since they would differ only by a term proportional to a constraint. The boundary term in (2.93) spoils this identification, however: as long as the vector ξ^i does not itself vanish at $\partial\Sigma$, we can no longer treat g_{kl} and g'_{kl} as being physically equivalent.

Put in slightly different terms, consider a transverse splitting of small fluctuations of a background metric g_{ij},

$$\delta g_{ij} = h_{ij} - (K\xi)_{ij}, \quad (K^\dagger h)_j = 0$$
(2.94)

with

$$(K\xi)_{ij} = {}^{(2)}\nabla_i\xi_j + {}^{(2)}\nabla_j\xi_i.$$
(2.95)

In the absence of a boundary, this splitting is unique, and provides a standard division of the metric into 'physical' and 'gauge' degrees of freedom [36, 298]. If Σ has a boundary, however, a unique decomposition requires boundary conditions that make $K^\dagger K$ self-adjoint. The simplest choice is

$$\xi^i\big|_{\partial\Sigma} = 0.$$
(2.96)

Just as in Chern–Simons theory, the 'would-be gauge' degrees of freedom $K\xi$ with $\xi^i \neq 0$ at $\partial\Sigma$ are potential new dynamical degrees of freedom at the boundary. Note, of course, that the degrees of freedom corresponding to Killing vectors – those ξ^i for which $(K\xi)_{ij} = 0$ – do *not* become dynamical, even if they are nonzero at $\partial\Sigma$, since they do not contribute to the right-hand side of (2.95). This observation will be important in our treatment of black hole statistical mechanics in chapter 12.

It is clear from equation (2.70) that these 'would-be diffeomorphism' degrees of freedom are related to the 'would-be gauge' degrees of freedom in the Chern–Simons formulation. Unfortunately, in the absence of a local decomposition of degrees of freedom in the metric formalism analogous to (2.81), it is not clear how to write down the metric equivalent of the boundary action (2.92). This means that for now, at least, the precise form of the boundary action for general relativity can only be given in 2+1 dimensions, since only then is a Chern–Simons formulation possible.

2.7 Comparing generators of invariances

In section 4, we computed the generators of diffeomorphisms in the ADM formulation and analyzed their Poisson algebra. In particular, we found that the algebra (2.47)–(2.48) was not a true Lie algebra – the commutators were characterized by field-dependent structure *functions* rather than structure constants. On the other hand, we have now seen that the first-order formalism allows us to express the diffeomorphisms as ordinary gauge transformations, whose generators should satisfy a genuine Lie algebra. By studying the relationship between these two formulations, we can gain further insight into the structure of (2+1)-dimensional gravity.

We begin by considering a decomposition of the first-order action (2.61) into space and time components. On a manifold with the topology $M \approx [0,1] \times \Sigma$, it is easy to check that up to boundary terms, the action can be written as

$$I = 2 \int dt \int_\Sigma d^2x \left\{ \epsilon^{ij} e_i{}^a \dot\omega_{ja} - \omega_t{}^a \mathscr{C}_a - e_t{}^a \tilde{\mathscr{C}}_a \right\}, \tag{2.97}$$

where the constraints are

$$\mathscr{C}^a[\omega, e] = \frac{1}{2} \epsilon^{ij} \left[\partial_i e_j{}^a - \partial_j e_i{}^a + \epsilon^{abc}(\omega_{ib} e_{jc} - \omega_{jb} e_{ic}) \right]$$

$$\tilde{\mathscr{C}}^a[\omega, e] = \frac{1}{2} \epsilon^{ij} \left[\partial_i \omega_j{}^a - \partial_j \omega_i{}^a + \epsilon^{abc}(\omega_{ib}\omega_{jc} - \Lambda e_{ib} e_{jc}) \right]. \tag{2.98}$$

These constraints have a straightforward intepretation. When $\Lambda = 0$, the $\tilde{\mathscr{C}}$ constraint tells us that $\omega_i{}^a$, the induced $SO(2,1)$ connection on Σ, is flat. The $\mathscr{C}$ constraint then implies that $e_i{}^a$ is a cotangent vector to the space of flat connections. Indeed, let $\omega_i{}^a(s)$ be a one-parameter family of flat connections – solutions of the $\tilde{\mathscr{C}}$ constraint – on Σ. Then the derivative of $\tilde{\mathscr{C}}^a[\omega(s)]$ with respect to s is

$$\frac{d}{ds}\tilde{\mathscr{C}}^a[\omega(s)] = \frac{1}{2}\epsilon^{ij}\left[\partial_i \frac{d\omega_j{}^a}{ds} - \partial_j \frac{d\omega_i{}^a}{ds} - \epsilon^{abc}\left(\omega_{ib}\frac{d\omega_{jc}}{ds} - \omega_{jb}\frac{d\omega_{jc}}{ds}\right)\right]$$

$$= \mathscr{C}^a\left[\omega(s), \frac{d\omega(s)}{ds}\right]. \tag{2.99}$$

This means that the $\mathscr{C}$ constraint is identically satisfied whenever e is a cotangent vector $d\omega/ds$ to the space of flat connections. Conversely, when the space of flat connections is sufficiently well-behaved, any triad satisfying $\mathscr{C}[\omega, e] = 0$ determines such a cotangent vector.

When $\Lambda \neq 0$, the geometric interpretation of the constraints is less transparent. In the Chern–Simons formulation, however, it is not too hard to show that the two constraints (2.99) are both contained in the field strength $\epsilon^{ij} F_{ij}$, and that the constraints again imply that the induced connection on Σ – now an $SO(2,1) \times SO(2,1)$ or $SL(2,\mathbb{C})$ connection – is flat.

From the action (2.97), we can read off the equal time Poisson brackets

$$\{e_i{}^a(x), \omega_{jb}(x')\} = -\frac{1}{2}\epsilon_{ij}\delta_b^a\delta^2(x - x'), \tag{2.100}$$

from which it is evident that

$$\{F[e, \omega], e_i{}^a\} = \frac{1}{2}\epsilon_{ij}\frac{\delta F}{\delta\omega_{ja}}$$

$$\{F[e, \omega], \omega_{ia}\} = \frac{1}{2}\epsilon_{ij}\frac{\delta F}{\delta e_j{}^a}, \tag{2.101}$$

the first-order analogs of (2.38). It follows from a simple computation that $\mathscr{C}^a$ and $\tilde{\mathscr{C}}^a$ generate the gauge transformations (2.66)–(2.67), and that (up to boundary terms, as usual) the generators satisfy the standard Lie algebra brackets

$$\left\{\mathscr{C}^a(x), \mathscr{C}^b(x')\right\} = -\frac{1}{2}\epsilon^{abc}\mathscr{C}_c(x)\delta^2(x - x')$$

$$\left\{\mathscr{C}^a(x), \tilde{\mathscr{C}}^b(x')\right\} = -\frac{1}{2}\epsilon^{abc}\tilde{\mathscr{C}}_c(x)\delta^2(x - x')$$

$$\left\{\tilde{\mathscr{C}}^a(x), \tilde{\mathscr{C}}^b(x')\right\} = \frac{\Lambda}{2}\epsilon^{abc}\mathscr{C}_c(x)\delta^2(x - x'). \tag{2.102}$$

Note that in contrast to the ADM formalism, the brackets are now those of a genuine Lie algebra. For $\Lambda = 0$, equation (2.102) may be recognized as a rescaled version of the commutators (2.73) for the Lie algebra of $ISO(2,1)$. Similarly, for $\Lambda \neq 0$, equation (2.102) gives the commutation relations for the Lie algebras of $SO(2,1) \times SO(2,1)$ (for $\Lambda < 0$) or $SL(2,\mathbb{C})$ (for $\Lambda > 0$).

As in section 4, we must introduce additional boundary terms to the constraints $\mathscr{C}^a$ and $\tilde{\mathscr{C}}^a$ if the spatial slice Σ is open. The analog of the generator $\mathscr{G}$ of equation (2.51) is now

$$G[\rho, \tau] = -\int_\Sigma d^2x \left(\tau_a\mathscr{C}^a + \rho_a\tilde{\mathscr{C}}^a\right), \tag{2.103}$$

and it is easy to see that the variation of $G[\rho, \tau]$ induces a boundary variation

$$\delta G = -2 \int_{\partial \Sigma} d\phi \left(\rho_a \delta \omega_{\parallel}{}^a + \tau_a \delta e_{\parallel}{}^a \right) + \ldots . \qquad (2.104)$$

As in the ADM case, we must cancel this expression by adding a new boundary term to G, in this case

$$\overline{G}[\bar{\rho}, \bar{\tau}] = 2 \int_{\partial \Sigma} d\phi \left(\bar{\rho}_a \omega_{\parallel}{}^a + \bar{\tau}_a e_{\parallel}{}^a \right). \qquad (2.105)$$

We shall see in the next chapter that this boundary term gives the same conserved charges – the total mass and angular momentum – as the corresponding ADM expression.

We can now express the ADM constraints $\mathcal{H}$ and $\mathcal{H}_i$ in terms of the first-order constraints $\mathcal{C}^a$ and $\tilde{\mathcal{C}}^a$. We should expect the results to be somewhat ambiguous: the first-order formalism has more symmetries (local Lorentz transformations as well as diffeomorphisms), and consequently more constraints, than the ADM formalism. To avoid this ambiguity, we shall derive the ADM constraints only modulo local Lorentz transformations; that is to say, we shall find a set of constraints that obey the algebra (2.47)–(2.48), but only up to terms proportional to the generator $\mathcal{C}^a$ of local $SO(2,1)$ transformations.

For simplicity, let us set the cosmological constant Λ to zero. We saw in the last section that a diffeomorphism generated by a vector ξ^i is equivalent to an $ISO(2,1)$ transformation with $\rho^a = \xi^i e_i{}^a$, plus a local Lorentz transformation, which we are ignoring. We might therefore expect that

$$\xi^i \mathcal{H}_i = 2\rho^a \tilde{\mathcal{C}}_a, \qquad (2.106)$$

or

$$\mathcal{H}_i = 2 e_i{}^a \tilde{\mathcal{C}}_a. \qquad (2.107)$$

The Hamiltonian constraint $\mathcal{H}$ is a bit harder to guess, but the answer is known from work on Ashtekar's formulation of (3+1)-dimensional gravity [11]:

$$\mathcal{H} = 2g^{-1/2} \epsilon_{abc} \epsilon^{ij} e_i{}^b e_j{}^c \tilde{\mathcal{C}}^a, \qquad (2.108)$$

where g is the determinant of the spatial metric,

$$g = \det(e_i{}^a e_{ja}). \qquad (2.109)$$

The form of this constraint can be made plausible by noting that

$$\epsilon_{abc} \epsilon^{ij} e_i{}^b e_j{}^c = \epsilon_{abc} \epsilon^{\mu\nu 0} e_\mu{}^b e_\nu{}^c = 2 e e^0{}_a = 2N\sqrt{g} e^0{}_a, \qquad (2.110)$$

so equation (2.108) is rather closely parallel to equation (2.15).

A tedious but straightforward calculation now shows that the generators (2.107) and (2.108) have the correct Poisson brackets (2.47)–(2.48), modulo terms proportional to $\mathscr{C}^a$. We have thus constructed the constraint algebra of general relativity from that of a gauge theory. The existence of such a construction was first noted by Bengtsson, and it provides yet another viewpoint from which to understand the simplicity of (2+1)-dimensional gravity: the difficult diffeomorphism constraints reduce to much simpler gauge constraints [35].

It is easy to see how the dimensionality of spacetime has entered this construction: we have made explicit use of the Levi–Civita tensor, whose form depends on the number of dimensions. In 3+1 dimensions, much of the simplicity found here is lost. Nevertheless, a related construction has been used to show that even in 3+1 dimensions, the phase space of general relativity 'embeds' in the phase space of a gauge theory; this is the structure that provides much of the power of Ashtekar's new variables [11].

3

A field guide to the (2+1)-dimensional spacetimes

In the last chapter, we investigated two formulations of the vacuum Einstein field equation in 2+1 dimensions. In this chapter, we will solve these field equations in several fairly simple settings, finding spacetimes that represent a collection of point particles, a rotating black hole, and a variety of closed universes with topologies of the form $[0,1] \times \Sigma$. In contrast to (3+1)-dimensional general relativity, where it is almost always necessary to impose strong symmetry requirements in order to find solutions, we shall see that for simple enough topologies, it is actually possible to find the *general* solution of the (2+1)-dimensional field equations.

The reader should be warned that this chapter is not a comprehensive survey of solutions of the (2+1)-dimensional field equations. In particular, I will spend a limited amount of time on the widely studied point particle solutions, and I will say little about solutions with extended ('string') sources and solutions in the presence of a nonvanishing matter stress–energy tensor. The latter are of particular interest for quantum theory – they offer models for studying the interaction of quantum gravity and quantum field theory – but systematic investigation of such solutions has only begun recently, and they are not yet very well understood.

3.1 Point sources

As a warm-up exercise, let us use the ADM formalism of chapter 2 to find the general stationary, axisymmetric solutions of the vacuum field equations with vanishing cosmological constant [90]. Such spacetimes are the (2+1)-dimensional analogs of the exterior Schwarzschild and Kerr metrics, representing the region outside a circularly symmetric gravitating source. These spacetimes have a possible (3+1)-dimensional interpretation as well: as described in chapter 1, the sources can be viewed as cross-sections of straight cosmic strings.

A general stationary, axisymmetric (2+1)-dimensional metric can be written in the form

$$ds^2 = -N(r)^2 dt^2 + f(r)^2 dr^2 + r^2 \left(d\phi - N^\phi(r) dt \right)^2, \qquad (3.1)$$

where I have used a spatial coordinate transformation to set the coefficient of $d\phi^2$ to be r^2. Physically, this coordinate choice means that a circle of constant r has circumference $2\pi r$. The spatial metric g_{ij} is simply

$$g_{ij} = \begin{pmatrix} f^2 & 0 \\ 0 & r^2 \end{pmatrix}, \qquad (3.2)$$

and an easy calculation shows that

$$^{(2)}R_{rr} = \frac{f'}{fr}$$

$$^{(2)}R_{\phi\phi} = \frac{rf'}{f^3}, \qquad (3.3)$$

and hence

$$\sqrt{^{(2)}g}\, ^{(2)}R = 2\frac{f'}{f^2}. \qquad (3.4)$$

Since the metric is static, the extrinsic curvature is

$$K_{ij} = -\frac{1}{2N}(^{(2)}\nabla_j N_i + {}^{(2)}\nabla_i N_j). \qquad (3.5)$$

The only nonvanishing component is thus

$$K_{r\phi} = -\frac{r^2}{2N}(N^\phi)', \qquad (3.6)$$

and the corresponding canonical momentum is

$$\pi_\phi{}^r = -\frac{r^3}{2Nf}(N^\phi)'. \qquad (3.7)$$

Let us first evaluate the momentum constraint (2.14),

$$^{(2)}\nabla_j \pi^{ij} = 0 = g^{il} \partial_k \pi^k{}_l - \frac{1}{2} g^{il} (\partial_l g_{jk}) \pi^{jk}. \qquad (3.8)$$

Since g_{jk} has only diagonal elements and π^{jk} is entirely off-diagonal, the last term of (3.8) vanishes. Hence $\pi_\phi{}^r$ is a constant,

$$\pi_\phi{}^r = A. \qquad (3.9)$$

We next consider the Hamiltonian constraint (2.13), which becomes

$$\frac{2}{\sqrt{^{(2)}g}} g_{\phi\phi} g_{rr} (\pi^{\phi r})^2 - \sqrt{^{(2)}g} \, {}^{(2)}R = \frac{2A^2 f}{r^3} - 2\frac{f'}{f^2} = 0. \qquad (3.10)$$

The solution of this equation is

$$\frac{1}{f^2} = B^2 + \frac{A^2}{r^2}, \qquad (3.11)$$

where B^2 is another integration constant, which must be positive to ensure that f^2 remains positive for large values of r.

To proceed further we shall also need one of the dynamical equations of motion coming from varying g_{ij} in the action (2.12). To find the appropriate equation, we substitute the ansatz (3.1) for the metric into the action* and vary f. The Hamiltonian constraint is

$$\mathcal{H} = 2fr(\pi^{\phi r})^2 - 2\frac{f'}{f^2}, \qquad (3.12)$$

the momentum constraint is independent of f, and all time derivatives vanish, so the action is

$$I_{\it eff} \sim - \int dt \int dr \left\{ 2Nrf(\pi^{\phi r})^2 - 2N\frac{f'}{f^2} \right\} + \text{terms independent of } f. \qquad (3.13)$$

The field equation obtained by varying f is thus

$$\frac{N'}{f^2} + Nr(\pi^{\phi r})^2 = 0. \qquad (3.14)$$

Combining (3.12) and (3.14) and the constraint $\mathcal{H} = 0$, we see that

$$\frac{N'}{N} = -\frac{f'}{f}, \qquad (3.15)$$

or

$$N = f^{-1} \qquad (3.16)$$

up to a constant factor that can be absorbed by a suitable rescaling of the time coordinate t.

---

* This procedure is a bit dangerous: by fixing an ansatz, we are restricting the possible variations of the metric, and hence finding only the metrics that are extrema within a limited class of geometries. In general we are not guaranteed that our solutions are extremal under variations that take the metric outside this class. Here, however, the ansatz is completely determined by symmetry considerations, and we should find any extrema with the appropriate symmetries.

To complete our solution, we can now use (3.7) to determine N^ϕ:

$$(N^\phi)' = -\frac{2Nf}{r^3}\pi_\phi{}^r = -\frac{2A}{r^3}, \tag{3.17}$$

so

$$N^\phi = C + \frac{A}{r^2}. \tag{3.18}$$

We may further restrict our attention to solutions for which $C = 0$, since otherwise the metric has unphysical asymptotic behavior.

Substituting (3.11), (3.16) and (3.18) into (3.1), we finally obtain a metric

$$ds^2 = -\left(B^2 + \frac{A^2}{r^2}\right)dt^2 + \left(B^2 + \frac{A^2}{r^2}\right)^{-1}dr^2 + r^2\left(d\phi + \frac{A}{r^2}dt\right)^2$$

$$= -\left(Bdt - \frac{A}{B}d\phi\right)^2 + \left(B^2 + \frac{A^2}{r^2}\right)^{-1}dr^2 + \left(r^2 + \frac{A^2}{B^2}\right)d\phi^2. \tag{3.19}$$

This metric can be put into a slightly more standard form by defining a new radial coordinate

$$\tilde{r} = \frac{1}{B^2}\left(A^2 + B^2r^2\right)^{1/2} \tag{3.20}$$

and by rescaling the time coordinate by a factor of B, to finally obtain

$$ds^2 = -(dt - \frac{A}{B}d\phi)^2 + d\tilde{r}^2 + B^2\tilde{r}^2d\phi^2. \tag{3.21}$$

To interpret the geometry of this spacetime, let us first consider the static case, $A = 0$. The line element (3.21) can then be made into the standard flat metric through the coordinate transformation

$$\phi \to \phi' = B\phi. \tag{3.22}$$

This is to be expected, since we saw in chapter 1 that the field equations require the curvature to vanish. Note, however, that ϕ' does not have the standard periodicity: as ϕ varies from 0 to 2π, ϕ' varies from 0 to $2\pi - \beta$, where

$$\beta = 2\pi(1 - B). \tag{3.23}$$

For $0 < \beta < 2\pi$, the spatial metric is actually that of a cone, obtained by cutting a wedge of opening angle β out of the plane and identifying the two edges (see figure 3.1). The angle β is known as the deficit angle

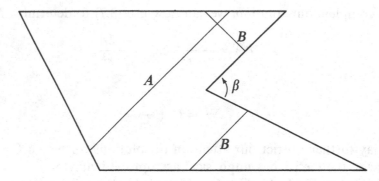

Fig. 3.1. A cone can be formed by cutting a wedge out of the plane and identifying the edges. The geodesics A and B are initially parallel, but eventually cross. (Note that B is really 'straight' and unbroken once the edge identifications are taken into account.)

of the cone. This is our first example of a theme that will be repeated throughout this book, the existence of nontrivial global geometry even in spacetimes whose curvature tensor vanishes.

When $A \neq 0$, a similar geometric description is possible. The standard flat metric can now be obtained by the coordinate transformations

$$\phi \to \phi' = B\phi$$
$$t \to t' = t - \frac{A}{B}\phi. \tag{3.24}$$

But the usual identification $(t, \phi) \sim (t, \phi + 2\pi)$ now becomes

$$(t', \phi') \sim (t' - 2\pi A/B, \phi' + 2\pi B). \tag{3.25}$$

In other words, we now start with flat Minkowski space, cut out a wedge with opening angle β, and identify the opposite edges with an extra time translation. It is easy to see that the resulting 'time-helical' structure leads to the presence of closed timelike curves.

To find a physical interpretation for the constants A and B, it is useful to examine the ADM equations of motion in the presence of sources, treating the conical singularity at $r = 0$ as a point particle. In the presence of matter, the field equations obtained from the variation of N and N^i become

$$\mathscr{H} = -\sqrt{{}^{(2)}g}\, T^0{}_0$$
$$\mathscr{H}_i = -\sqrt{{}^{(2)}g}\, T^0{}_i, \tag{3.26}$$

and the mass of an isolated source is thus

$$m = \int d^2x \sqrt{{}^{(2)}g}\, T^0{}_0 = -\int d^2x\, \mathscr{H}. \tag{3.27}$$

The only term in the Hamiltonian constraint (3.10) that has a chance of behaving peculiarly at $r = 0$ is the spatial curvature $^{(2)}R$. Indeed, recall from section 4 of chapter 2 that the curvature can be written in the form

$$\int_\Sigma d^2x \sqrt{^{(2)}g}\,^{(2)}R = \int_{\partial\Sigma} d\phi\, v^\perp = 2\pi v^\perp, \qquad (3.28)$$

where it is evident from (3.4) that

$$v^\perp = -\frac{2}{f} + \text{const.} \sim -2B + \text{const.} \qquad (3.29)$$

as $r \to \infty$. We can fix the constant by noting that when $B = 1$ and $A = 0$, the metric g_{ij} is that of flat Euclidean two-space, for which the integral (3.28) must vanish. Hence

$$v^\perp = 2 - 2B = \frac{\beta}{\pi}, \qquad (3.30)$$

and the total curvature integral is 2β. Restoring factors of G, equation (3.27) thus becomes

$$m = \frac{1}{16\pi G}\int d^2x \sqrt{^{(2)}g}\,^{(2)}R = \frac{\beta}{8\pi G}. \qquad (3.31)$$

A similar analysis can be applied to the angular momentum of the source, which in asymptotically Cartesian coordinates is

$$J^{ij} = \int d^2x \left(x^i T^{0j} - x^j T^{0i} \right) = 2\int_{\partial\Sigma} d\phi \left(x^i \pi^{\perp j} - x^j \pi^{\perp i} \right), \qquad (3.32)$$

where I have used the momentum constraints to write T^{0i} in terms of π^{ij}. Setting

$$\pi^{\perp i} = n_j \pi^{j\phi}\partial_\phi x^i = -\frac{A}{Br^2}\epsilon^{ik}x^k \qquad (3.33)$$

near infinity, we see that

$$J^{ij} = \frac{1}{4G}\frac{A}{B}\epsilon^{ij}, \qquad (3.34)$$

so A/B is a measure of the angular momentum of the source.

The reader may be uncomfortable with these arguments, based as they are on an analysis that is at least superficially noncovariant. Fortunately, the interpretation of the integration constants A and B may be checked in a manifestly covariant manner by calculating the conserved charges of chapter 2 associated with rotations and time translations at infinity. We have already computed the quantity $v^\perp$ appearing in (2.56); combining this result with the expression (3.9) for the canonical momentum, we see that the total boundary diffeomorphism generator is

$$\bar{\mathscr{G}}[\bar{\xi}, \bar{\xi}^i] = \frac{1}{16\pi G} \int_{\partial\Sigma} d\phi \left[(2 - 2B)\bar{\xi} + 2A\bar{\xi}^\phi\right], \tag{3.35}$$

where I have again restored the constants.

In particular, a diffeomorphism that is asymptotically a time translation has $\bar{\xi} = 1$, giving $\bar{\mathscr{G}} = \beta/8\pi G$. We thus confirm that β is the conserved quantity associated with time translations at spatial infinity, that is, the mass. Similarly, an asymptotic rotation is described by $\bar{\xi}^\phi = 1$ and $\bar{\xi} = A/B$. (The time translation component reflects the 'time-helical' structure of the metric, and is chosen so that the coordinate t' in (3.24) is left invariant.) The associated charge is then $\bar{\mathscr{G}} = A/4GB$, which is thus the angular momentum, the conserved quantity associated with rotations at infinity. Equivalent expressions for the mass and angular momentum may be found by computing the quasilocal mass of Brown and York associated with a surface at infinity [43].

The mass and angular momentum, constructed as integrals at spatial infinity, are the (2+1)-dimensional analogs of the ADM mass and angular momentum in standard general relativity. Similar integrals exist for any isolated system of sources. Note, however, that because of the asymptotically conical structure of our solutions, there are no asymptotic symmetries representing spatial translations or boosts, and thus no analog of the full ADM momentum vector. One way to understand this is to observe that while the metric (3.21) admits *local* solutions of the Killing equation corresponding to the full Poincaré group, most of the resulting would-be Killing vectors are not preserved under the identifications (3.25); only the Killing vectors corresponding to rotations and time translations are globally defined.

It is also instructive to examine the conserved charges in the first-order formalism. A suitable triad for the metric (3.21) is

$$
\begin{aligned}
e^0 &= dt - (A/B)d\phi \\
e^1 &= d\tilde{r} \\
e^2 &= B\tilde{r}\, d\phi,
\end{aligned}
\tag{3.36}
$$

and it is easily checked that the only nonvanishing component of the spin connection (2.63) is

$$\omega^0 = -Bd\phi, \tag{3.37}$$

for which the field equations (2.62) and (2.65) are clearly satisfied. There are now two candidates for asymptotic symmetries, that is, gauge transformations that leave the frame and spin connection fixed at infinity:

$$\rho^0 \sim 1, \tag{3.38}$$

corresponding to a time translation, and

$$\tau^0 \sim -1, \qquad \rho^2 \sim \tilde{r}, \tag{3.39}$$

corresponding to a spatial rotation. For the time translation (3.38), the generator (2.105) is $\overline{G} = 2\beta - 4\pi$, in agreement with the ADM result up to an overall additive constant. Similarly, for the rotation (3.39), we find $\overline{G} = 4\pi A/B$, again coinciding with the ADM result.

The analog of the metric (3.21) for a system of stationary, spinning sources is also known [78]. In 'Cartesian' coordinates, the metric for a collection of point particles with masses M_α and spins J_α located at positions $\mathbf{a}_\alpha$ is

$$ds^2 = \left(dt + 4G \sum_\alpha J_\alpha \frac{\epsilon_{ij}(x^i - a^i_\alpha)dx^j}{|\mathbf{r} - \mathbf{a}_\alpha|^2} \right)^2 - \prod_\alpha |\mathbf{r} - \mathbf{a}_\alpha|^{-8GM_\alpha} d\mathbf{r} \cdot d\mathbf{r}. \tag{3.40}$$

In contrast to the (3+1)-dimensional case, this is an *exact* stationary solution for multiple sources. The existence of such a solution reflects the peculiar weak-field limit discussed in chapter 1: stationary sources experience no Newtonian interaction. Moving sources, on the other hand, experience nontrivial scattering, as one would expect from the conical nature of the metric. Figure 3.1, for example, shows two initially parallel geodesics on a conical background; it is evident that they are 'scattered' as they pass the apex of the cone, intersecting each other at the far side. There has been considerable work on the structure of spacetimes representing moving sources in 2+1 dimensions, which I will not discuss here; see, for example, reference [34] for an interesting treatment.

3.2 The (2+1)-dimensional black hole

The axially symmetric spacetimes of the last section become even more interesting in the presence of a negative cosmological constant $\Lambda = -1/\ell^2$, for which spacetime is asymptotically anti-de Sitter. Most of the preceding derivation can be repeated, with a few changes:

1. The Hamiltonian constraint (3.10) is now

$$\frac{2A^2 f}{r^3} - 2\frac{f'}{f^2} - \frac{2r}{\ell^2} f = 0, \tag{3.41}$$

which has as its solution

$$\frac{1}{f^2} = B^2 + \frac{A^2}{r^2} + \frac{r^2}{\ell^2}. \tag{3.42}$$

The integration constant B^2 need no longer be positive, however, since the large r behavior of f^2 is now controlled by the term r^2/ℓ^2.

2. The equation of motion (3.14) for N now becomes

$$\frac{N'}{f^2} + Nr(\pi^{\phi r})^2 - \frac{Nr}{\ell^2} = 0. \tag{3.43}$$

The solution, however, is still $N = f^{-1}$.

Renaming some of the constants, we obtain a metric

$$ds^2 = -N^2 dt^2 + r^2 \left(d\phi^2 + N^\phi dt \right)^2 + N^{-2} dr^2 \tag{3.44}$$

with

$$N^2 = -M + \frac{r^2}{\ell^2} + \frac{J^2}{4r^2}, \qquad N^\phi = -\frac{J}{2r^2}. \tag{3.45}$$

This spacetime is the (2+1)-dimensional black hole of Bañados, Teitelboim, and Zanelli (BTZ) [23, 22]. It has an event horizon at $r = r_+$ and an inner horizon at r_-, where

$$r_\pm^2 = \frac{\ell^2}{2} \left[M \pm \left(M^2 - \frac{J^2}{\ell^2} \right)^{1/2} \right] \tag{3.46}$$

are the zeros of the lapse function N. That $r = r_+$ is a genuine event horizon is most easily seen by changing to Eddington–Finkelstein coordinates,

$$dv = dt + \frac{1}{N^2} dr, \quad d\tilde\phi = d\phi - \frac{N^\phi}{N^2} dr, \tag{3.47}$$

in which the metric becomes

$$ds^2 = -N^2 dv^2 + 2 dv dr + r^2 \left(d\tilde\phi + N^\phi dv \right)^2. \tag{3.48}$$

It is now evident that the surface $r = r_+$ is a null surface, generated by the geodesics

$$r(\lambda) = r_+, \quad \frac{d\tilde\phi}{d\lambda} + N^\phi(r_+)\frac{dv}{d\lambda} = 0. \tag{3.49}$$

Moreover, $r = r_+$ is a marginally trapped surface: at $r = r_+$, any null geodesic satisfies

$$\frac{dv}{d\lambda}\frac{dr}{d\lambda} = -\frac{r_+^2}{2}\left(\frac{d\tilde{\phi}}{d\lambda} + N^\phi(r_+)\frac{dv}{d\lambda}\right)^2 \leq 0, \qquad (3.50)$$

so r decreases or, for the geodesics (3.49), remains constant as v increases.

Like the outer horizon of the Kerr metric, the surface $r = r_+$ is also a Killing horizon, that is, a null surface to which a Killing vector is normal. The relevant Killing vector is

$$\chi = \partial_v - N^\phi(r_+)\partial_{\tilde{\phi}}. \qquad (3.51)$$

Given such a Killing vector, the surface gravity κ is defined by

$$\kappa^2 = -\frac{1}{2}\nabla^a\chi^b\nabla_a\chi_b, \qquad (3.52)$$

which may be easily computed:

$$\kappa = \frac{r_+^2 - r_-^2}{\ell^2 r_+}. \qquad (3.53)$$

This quantity can be interpreted as the red-shifted proper acceleration of a zero angular momentum observer at the horizon. We shall see in chapter 12 that κ plays an important role in black hole thermodynamics.

The (2+1)-dimensional black hole may also be expressed in Kruskal-like coordinates. We define new null coordinates

$$u = \rho(r)e^{-at}, \qquad v = \rho(r)e^{at}, \qquad \text{with} \quad \frac{d\rho}{dr} = \frac{a\rho}{N^2}. \qquad (3.54)$$

As in the case of the Kerr metric, we need two patches, $r_- < r < \infty$ and $0 < r < r_+$, to cover the black hole spacetime. In each patch, the metric (3.44) takes the form

$$ds^2 = \Omega^2 dudv + r^2(d\tilde{\phi} + N^\phi dt)^2 \qquad (3.55)$$

where

$$\Omega_+^2 = \frac{(r^2 - r_-^2)(r + r_+)^2}{a_+^2 r^2 \ell^2}\left(\frac{r - r_-}{r + r_-}\right)^{r_-/r_+},$$

$$\tilde{\phi}_+ = \phi + N^\phi(r_+)t, \qquad a_+ = \frac{r_+^2 - r_-^2}{\ell^2 r_+} \qquad (r_- < r < \infty)$$

$$\Omega_-^2 = \frac{(r_+^2 - r^2)(r + r_-)^2}{a_-^2 r^2 \ell^2}\left(\frac{r_+ - r}{r_+ + r}\right)^{r_+/r_-},$$

$$\tilde{\phi}_- = \phi + N^\phi(r_-)t, \qquad a_- = \frac{r_-^2 - r_+^2}{\ell^2 r_-} \qquad (0 < r < r_+)$$

$$(3.56)$$

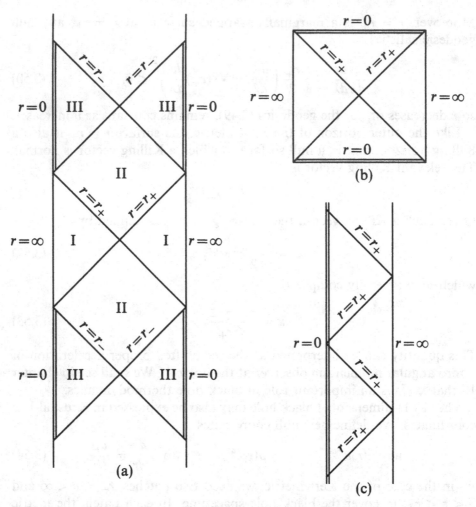

Fig. 3.2.   The (2+1)-dimensional black hole is characterized by a different Penrose diagram for (a) the generic case; (b) the $J = 0$ case; and (c) the extremal $(J = \pm M\ell)$ case.

with r and t viewed as implicit functions of u and v. (The explicit coordinate transformations to these Kruskal-like coordinates are given in reference [22].)

As in the case of the Kerr black hole, an infinite number of such Kruskal patches may be fit together to form a maximal solution, whose Penrose diagram is shown in diagram (a) of figure 3.2. This diagram differs from that of the Kerr metric at $r = \infty$, reflecting the fact that the BTZ black hole is asymptotically anti-de Sitter rather than asymptotically flat, but the overall structure is similar. In particular, it is evident that $r = r_+$ is an event horizon, while the inner horizon $r = r_-$ is a Cauchy horizon for region I. When $J = 0$, the Penrose diagram collapses to that of diagram

(b) of figure 3.2, which is similar in structure, except for its asymptotic behavior, to the diagram for the ordinary Schwarzschild solution. For the extreme case, $J = \pm M\ell$, the Penrose diagram is that of diagram (c) of figure 3.2.

As the notation suggests, M and J are the mass and angular momentum of the black hole, in the units $8G = 1$ of reference [23]. To see this, we can repeat the derivation of the conserved charges in section 1, with suitable adjustments to account for the fact that the black hole spacetime is not asymptotically flat. First, in an asymptotically anti-de Sitter spacetime, the boundary diffeomorphism generator $\bar{\mathscr{G}}$ must be normalized by subtracting a 'background' value, which we can take to be its value at $M = J = 0$.[†] The quantity $v^{\perp}$ in equation (3.29) is thus

$$v^{\perp} = -\frac{2}{f} + \frac{2}{f}\bigg|_{M=J=0} = -2\left(-M + \frac{r^2}{\ell^2} + \frac{J^2}{4r^2}\right)^{1/2} + \frac{2r}{\ell} \sim \frac{M\ell}{r}. \tag{3.57}$$

Moreover, the time–time component of the metric is now asymptotically $g_{tt} \sim r^2/\ell^2$, so a unit translation in coordinate time t corresponds to a vector

$$\bar{\chi} = n_{\mu}\chi^{\mu} \sim \frac{r}{\ell} \tag{3.58}$$

(where n^{μ} is a unit vector), rather than $\bar{\chi} \sim 1$. Inserting these expressions into equation (2.56) for the generator of asymptotic time translations, we find a total mass of $M/8G$, as claimed. This mass is to be interpreted as the generator of translations in the 'Killing time' t, the time coordinate that parametrizes the orbits of the Killing vector for which the black hole metric is stationary. The absolute normalization of t, and thus of M, may be fixed by requiring that the metric approach a standard anti-de Sitter form at spatial infinity. A similar argument shows that J is again the angular momentum.

The BTZ metric (3.44) has constant negative curvature, and like the point sources considered above, it can be represented as a region of a standard constant curvature space – in this case, anti-de Sitter space – with appropriate identifications of boundaries. One way to see this is to

---

[†] The precise choice of subtraction is somewhat arbitrary; a change would have the effect of shifting the location of the zero of the mass scale.

carry out yet another coordinate transformation,

$$x = \left(\frac{r^2 - r_+^2}{r^2 - r_-^2}\right)^{1/2} \cosh\left(\frac{r_+}{\ell^2}t - \frac{r_-}{\ell}\phi\right) \exp\left\{\frac{r_+}{\ell}\phi - \frac{r_-}{\ell^2}t\right\}$$

$$y = \left(\frac{r^2 - r_+^2}{r^2 - r_-^2}\right)^{1/2} \sinh\left(\frac{r_+}{\ell^2}t - \frac{r_-}{\ell}\phi\right) \exp\left\{\frac{r_+}{\ell}\phi - \frac{r_-}{\ell^2}t\right\}$$

$$z = \left(\frac{r_+^2 - r_-^2}{r^2 - r_-^2}\right)^{1/2} \exp\left\{\frac{r_+}{\ell}\phi - \frac{r_-}{\ell^2}t\right\}, \qquad (3.59)$$

for which the metric in the patch $r > r_+$ becomes

$$ds^2 = \frac{\ell^2}{z^2}(dx^2 - dy^2 + dz^2) \qquad (z > 0). \qquad (3.60)$$

This expression may be recognized as the standard 'upper half-space' representation of constant negative curvature three-space; its Euclidean version, formed by transforming y to iy, is the conventional metric for the hyperbolic space $\mathbf{H}^3$. Note, however, that the black hole metric does not cover the entire upper half-space: periodicity in the 'Schwarzschild' angular coordinate ϕ requires that we identify

$$(x, y, z) \sim$$

$$e^{2\pi r_+/\ell}\left(x\cosh\frac{2\pi r_-}{\ell} - y\sinh\frac{2\pi r_-}{\ell}, \; y\cosh\frac{2\pi r_-}{\ell} - x\sinh\frac{2\pi r_-}{\ell}, \; z\right). \quad (3.61)$$

The BTZ black hole can thus be obtained as a region of anti-de Sitter space with appropriate identifications of boundaries. This global geometry and its physical implications will be treated in more detail in chapter 12.

3.3 The torus universe

For our next collection of (2+1)-dimensional spacetimes, we turn to closed 'cosmological' solutions with the spatial topology of a torus, that is, spacetimes with the topology $[0, 1] \times T^2$ in which the T^2 slices are spacelike. These solutions may be found most directly via the ADM formalism developed in chapter 2; in particular, the York time-slicing $TrK = -T$ greatly simplifies many of the computations. Following the methods of chapter 2, we shall start by examining the physical phase space, parametrized by a set of moduli and their conjugate momenta.

The moduli space $\mathcal{N}(T^2)$ of a torus – the space of flat metrics of unit area – is parametrized by a single complex number $\tau = \tau_1 + i\tau_2$, the modulus, where we can assume without loss of generality that $\tau_2 >$

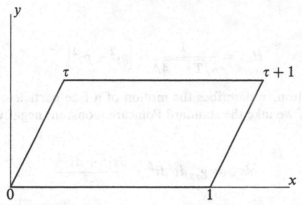

Fig. 3.3. A torus of modulus τ can be represented as a parallelogram with opposite sides identified.

0. (Readers unfamiliar with this moduli space may wish to consult appendix A.) For a given modulus, the corresponding flat metric is

$$\tilde{g}_{ij}dx^i dx^j = \tau_2^{-1}|dx + \tau dy|^2, \qquad (3.62)$$

where x and y are periodic coordinates with period one. Geometrically, a torus with modulus τ may be represented as a parallelogram on the complex plane, with vertices 0, 1, τ, and $\tau + 1$, appropriately rescaled to have unit area, as shown in figure 3.3.

As we saw in chapter 2, the variable conjugate to the flat metric $\tilde{g}_{ij}$ is a transverse traceless tensor p^{ij}, which for the torus may be parametrized as

$$p^{ij} = \frac{1}{2}\begin{pmatrix} (\tau_1{}^2 - \tau_2{}^2)p_2 - 2\tau_1\tau_2 p_1 & \tau_2 p_1 - \tau_1 p_2 \\ \tau_2 p_1 - \tau_1 p_2 & p_2 \end{pmatrix}. \qquad (3.63)$$

The dependence on parameters has been chosen so that p_1 and p_2 are conjugate to τ_1 and τ_2, i.e.,

$$\delta p^{ij} \wedge \delta g_{ij} = \delta p_1 \wedge \delta \tau_1 + \delta p_2 \wedge \delta \tau_2. \qquad (3.64)$$

Together, τ and $p = p_1 + ip_2$ constitute the initial data in the York time-slicing, and their evolution is determined by the Hamiltonian (2.37). To find this Hamiltonian, we must solve the elliptic differential equation (2.35) for the scale factor λ. For the torus topology – and uniquely for this case – the coefficients in (2.35) are spatial constants, and λ can be found algebraically:

$$e^{4\lambda} = \frac{2}{T^2 - 4\Lambda}\tilde{g}_{ij}\tilde{g}_{kl}p^{ik}p^{jl} = \frac{1}{T^2 - 4\Lambda}\tau_2{}^2\left(p_1{}^2 + p_2{}^2\right), \qquad (3.65)$$

and thus

$$H_{red} = \frac{1}{\sqrt{T^2 - 4\Lambda}} \tau_2 \left[p_1{}^2 + p_2{}^2 \right]^{1/2}. \tag{3.66}$$

This Hamiltonian describes the motion of a free particle on the hyperbolic plane: if we take the standard Poincaré (constant negative curvature) metric

$$ds^2_{\mathcal{N}} = g^{\mathcal{N}}_{\alpha\beta} d\tau^{\alpha} d\tau^{\beta} = \frac{d\tau_1{}^2 + d\tau_2{}^2}{\tau_2{}^2} \tag{3.67}$$

on the upper half-plane $\tau_2 > 0$, H_{red} is of the form $(g^{\mathcal{N}\alpha\beta} p_{\alpha} p_{\beta})^{1/2}$. The resulting dynamics is fairly straightforward. If we define a new time coordinate t by

$$T = e^t + \Lambda e^{-t}, \qquad dt = \frac{dT}{(T^2 - 4\Lambda)^{1/2}}, \tag{3.68}$$

the equations of motion coming from the reduced phase space action are

$$\frac{d\tau_1}{dt} = \tau_2 \left(p_1{}^2 + p_2{}^2 \right)^{-1/2} p_1, \qquad \frac{dp_1}{dt} = 0,$$
$$\frac{d\tau_2}{dt} = \tau_2 \left(p_1{}^2 + p_2{}^2 \right)^{-1/2} p_2, \qquad \frac{dp_2}{dt} = -\left(p_1{}^2 + p_2{}^2 \right)^{-1/2}. \tag{3.69}$$

The general solution is easily found:

$$\tau_1 = \beta + \alpha \tanh(t - t_0)$$
$$\tau_2 = \alpha \operatorname{sech}(t - t_0)$$
$$p_1 = \text{const.}$$
$$p_2 = -p_1 \sinh(t - t_0), \tag{3.70}$$

where t_0, α, β, and p_1 are integration constants. The resulting motion is a semicircle in the upper half-plane centered on the real axis,

$$(\tau_1 - \beta)^2 + \tau_2{}^2 = \alpha^2, \tag{3.71}$$

and it is straightforward to check that this trajectory is a geodesic with respect to the metric (3.67).

The spacetimes given by equation (3.70) are clearly dynamical: the spatial geometry, as described by the modulus τ, changes from slice to slice. In addition, the area changes with time. As noted earlier, the spatial area of a constant T slice is given by the Hamiltonian (3.66) evaluated on the solution (3.70):

$$H_{red} = \frac{\alpha p_1}{(T^2 - 4\Lambda)^{1/2}}.$$ (3.72)

For $\Lambda \geq 0$, the solutions display a 'big bang' or 'big crunch' singularity: as T increases from an initial value of $2\sqrt{\Lambda}$, the universe collapses, reaching a final singularity of zero area at $T = \infty$. (The time-reversed solution represents an expanding universe.) For Λ negative, on the other hand, the area reaches a minimum value: the universe 'bounces'.

Note that as T approaches infinity, τ_2 also goes to zero. This indicates that as the universe approaches a big bang or big crunch, the spatial geometry also becomes singular: the parallelogram of figure 3.3 collapses to a line. Similar – but typically much more complex – singular behavior of the spatial geometry is known to occur for (2+1)-dimensional universes with more complicated topologies; a good deal of the mathematics is understood, in a fairly technical form, but this behavior has not yet been analyzed in the physics literature [195].

So far, I have been rather cavalier in my treatment of the symmetries of these solutions. By construction, the solutions are invariant under the 'small' spacetime diffeomorphisms, that is, the diffeomorphisms that can be built from infinitesimal transformations. The torus also admits 'large' diffeomorphisms, however, generated by Dehn twists around the circumferences. These act nontrivially on the moduli τ and the momenta p. As described in appendix A, the group of such large diffeomorphisms of the torus (modulo 'small' diffeomorphisms) has two generators,

$$S : \tau \to -\frac{1}{\tau}, \qquad p \to \bar{\tau}^2 p$$
$$T : \tau \to \tau + 1, \qquad p \to p.$$ (3.73)

This means that not all moduli are geometrically distinct. Rather, the independent geometries are characterized by values of τ that range over a single fundamental region, such as the 'keyhole' region shown in figures 3.4 and A.6. Equation (3.71) for the trajectory is therefore somewhat misleading; projected back to a single fundamental region, a portion of a typical trajectory looks like the path shown in figure 3.4. The resulting motion is quite complicated: there are arbitrarily long closed geodesics, and the geodesic flow is actually ergodic [80, 150]. Note, though, that the evolution of the physical geometry of space is *not* ergodic. The full geometry depends on both the modulus and the area, and the time-dependence (3.72) of the area is simple and well-behaved.

The role of large diffeomorphisms is a subtle one, but it will be important later when we discuss quantization. For the torus, the upper

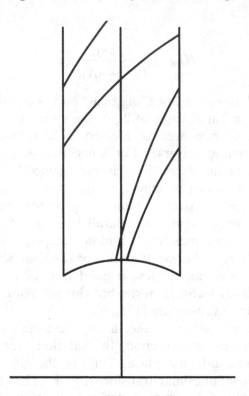

Fig. 3.4. A standard representation of the torus moduli space is given by a 'keyhole' region of Teichmüller space; opposite edges are identified, and points on the bottom arc are identified with their reflections through the y axis. A portion of a geodesic is shown.

half-plane $\{\tau : \tau_2 > 0\}$ is known as the *Teichmüller space*, while the quotient of Teichmüller space by the group action (3.73) is known as *moduli space*. More generally, the Teichmüller space $\tilde{\mathcal{N}}(\Sigma)$ of a surface Σ is the space of constant curvature metrics on Σ modulo small diffeomorphisms, while the moduli space $\mathcal{N}(\Sigma)$ is the space of constant curvature metrics modulo *all* diffeomorphisms. In classical (2+1)-dimensional gravity, it is the moduli space that parametrizes inequivalent spacetimes; the translation of this condition to the quantum theory will provide valuable constraints.

The reduced phase space variables τ and p describe the physical degrees of freedom, but we have not yet found the full spacetime metric $g_{\mu\nu}$. In particular, we do not yet have expressions for the lapse and shift functions. To proceed further, it is useful to switch to the first-order formalism of chapter 2. Note first that from equations (3.62) and (3.70) and equation

(2.21) of chapter 2, the spatial line element $d\sigma^2 = g_{ij}dx^i dx^j$ is

$$d\sigma^2 = e^{2\lambda}\tilde{g}_{ij}dx^i dx^j$$

$$= \frac{\left(p_1{}^2 + p_2{}^2\right)^{1/2}}{\left(T^2 - 4\Lambda\right)^{1/2}}\left[dx^2 + 2\tau_1 dx dy + (\tau_1{}^2 + \tau_2{}^2)dy^2\right]$$

$$= \frac{1}{1 - \Lambda e^{-2t}}\left[(adx + bdy)^2 + e^{-2t}(\lambda dx + \mu dy)^2\right], \qquad (3.74)$$

where I have used the abbreviations

$$a = \sqrt{\frac{p_1}{2}}e^{-t_0/2}, \qquad b = \sqrt{\frac{p_1}{2}}e^{-t_0/2}(\beta + \alpha),$$

$$\lambda = -\sqrt{\frac{p_1}{2}}e^{t_0/2}, \qquad \mu = -\sqrt{\frac{p_1}{2}}e^{t_0/2}(\beta - \alpha). \qquad (3.75)$$

This 'factorized' form makes it easy to write down the triad e^a:

$$e^0 = N(t)dt$$
$$e^1 = Fe^{t/2}(adx + bdy)$$
$$e^2 = Fe^{-t/2}(\lambda dx + \mu dy), \qquad (3.76)$$

where

$$F^2 = \frac{1}{e^t - \Lambda e^{-t}} = \left(T^2 - 4\Lambda\right)^{-1/2}, \qquad (3.77)$$

and the lapse function $N(t)$ is still an unknown function of time. It is now easy to check that the torsion constraints (2.62) are satisfied by

$$\omega^0 = 0$$

$$\omega^1 = -\frac{F^2}{N}e^t e^2$$

$$\omega^2 = \frac{F^2\Lambda}{N}e^{-t}e^1. \qquad (3.78)$$

The remaining field equations (2.65) are now completely straightforward, and yield

$$N = F^2. \qquad (3.79)$$

The full spacetime metric is thus

$$ds^2 = -F^4 dt^2 + F^2 e^t(adx + bdy)^2 + F^2 e^{-t}(\lambda dx + \mu dy)^2. \qquad (3.80)$$

Related expressions have appeared in references [52, 106, 119, 153, 182].

We will later need expressions for the moduli τ_α and the momenta p_α in terms of the coefficients a, b, λ, and μ of equation (3.75). A simple computation gives

$$\tau = \tau_1 + i\tau_2 = \frac{b + i\mu e^{-t}}{a + i\lambda e^{-t}}$$
$$p = p_1 + ip_2 = -ie^t \left(a - i\lambda e^{-t}\right)^2$$
$$H_{red} = F^2(\mu a - \lambda b). \tag{3.81}$$

The symplectic structure on the space of parameters a, b, λ, and μ can then be read off from equation (3.81):

$$\delta p_1 \wedge \delta\tau_1 + \delta p_2 \wedge \delta\tau_2 = \frac{1}{2}(\delta\bar{p} \wedge \delta\tau + \delta p \wedge \delta\bar{\tau}) = 2(\delta b \wedge \delta\lambda - \delta a \wedge \delta\mu). \tag{3.82}$$

Thus (μ, a) and (λ, b) are conjugate pairs.

In addition to the four-parameter family of spacetimes (3.80), a further family of static spacetimes with the topology $[0, 1] \times T^2$ can be found when $\Lambda = 0$. These spacetimes are characterized by vanishing extrinsic curvature, $K_{ij} = 0$, so the York time-slicing $TrK = -T$ breaks down. On the other hand, the Hamiltonian constraint (2.28) is now trivial: it simply requires that λ be a constant. The resulting metrics are

$$ds^2 = -dt^2 + \frac{V^{1/2}}{\tau_2} |dx + \tau dy|^2, \tag{3.83}$$

where x and y are again periodic with period 1, τ is now time-independent, and V is a (constant) spatial area. In contrast to the spacetimes (3.80), these geometries are nondynamical, describing completely time-independent torus universes. This set of spacetimes is rather pathological from the point of view of quantum gravity – the symplectic form analogous to (3.82) is degenerate, and there are no natural canonical commutators. The metrics (3.83) are often ignored when dealing with the quantum theory, but they may be important for determining boundary conditions for wave functions.

Yet another four-parameter family of solutions has been found through the first-order formalism. These spacetimes again have the global topology $[0, 1] \times T^2$, but the T^2 slices are not spacelike, so the corresponding metrics do not appear in the ADM formalism. Such solutions necessarily contain closed timelike curves, and their relevance to quantum gravity is not clear. Louko and Marolf have carefully analyzed the full set of solutions of the first-order field equations with the topology $[0, 1] \times T^2$, and have shown that the space of solutions is quite badly behaved: in fact, it is not even

Hausdorff [108, 181]. This makes a quantum theory based on this full space of solutions rather problematical.

Like the point particle and black hole metrics, the torus geometries given by (3.80) and (3.83) can be formed by making appropriate identifications of points in a constant curvature space. This process is both rather complicated and extremely important; its discussion will be postponed until the next chapter.

3.4 Other topologies

The torus universes $[0, 1] \times T^2$ described in the last section provide a useful collection of $(2+1)$-dimensional spacetimes. The torus topology is rather atypical, however – the torus admits a flat metric, for instance, and has an abelian fundamental group – and it is natural to look for generalizations. Unfortunately, much less is known about spacetimes with the topology $[0, 1] \times \Sigma$ when Σ is a surface of genus $g \neq 1$.

Let us begin with the simplest case, the spherical universe $[0, 1] \times S^2$. It is a well-known result of Riemann surface theory that the two-sphere has no moduli – S^2 has a unique constant curvature metric with $k = 1$ – and that the sphere admits no transverse traceless tensors p^{ij}. The ADM metric is therefore determined entirely by the conformal factor λ, which by (2.35) must satisfy

$$\tilde{\Delta}\lambda - \frac{1}{4}(T^2 - 4\Lambda)e^{2\lambda} - \frac{1}{2} = 0. \tag{3.84}$$

Integrating over the two-sphere, we see that

$$\int_{S^2} d^2x \sqrt{\tilde{g}} \left[\frac{1}{4}(T^2 - 4\Lambda)e^{2\lambda} + \frac{1}{2} \right] = 0. \tag{3.85}$$

If $\Lambda \leq 0$, both terms in the integrand are positive, and this equation has no solution. If $\Lambda > 0$, there is at least one solution,

$$e^{2\lambda} = \frac{2}{4\Lambda - T^2}. \tag{3.86}$$

This is, in fact, the unique solution, corresponding to the de Sitter metric

$$ds^2 = \frac{4}{4\Lambda - T^2} \left(-\frac{dT^2}{4\Lambda - T^2} + d\theta^2 + \sin^2\theta d\phi^2 \right). \tag{3.87}$$

This spacetime is not very interesting from the point of view of quantum gravity, since it has no gravitational degrees of freedom, but it could be significant if matter couplings were added.

If we drop the requirement of orientability, we can also consider the case of a Klein bottle universe $[0, 1] \times K$. This topology has been analyzed by Louko, who shows that it has a reduced phase space that is, roughly speaking, half of the $[0, 1] \times T^2$ phase space described in the preceding section [180]. This is not a coincidence: the torus is the double cover of the Klein bottle, and metrics on the Klein bottle can be pulled back to metrics on the torus with appropriate symmetries.

We next consider spacetimes $[0, 1] \times \Sigma$ in which Σ is a surface of genus $g > 1$. Let us assume for simplicity that the cosmological constant vanishes. Then one set of solutions is quite easy to find: if we choose initial values $p^{ij} = 0$, equation (2.35) becomes

$$\tilde{\Delta}\lambda - \frac{T^2}{4}e^{2\lambda} + \frac{1}{2} = 0, \tag{3.88}$$

which is solved by

$$e^{2\lambda} = \frac{2}{T^2}. \tag{3.89}$$

The Hamiltonian is thus a constant, and the momenta remain zero. This procedure generates a $(6g - 6)$-parameter family of solutions: we can choose any set of moduli on an initial slice, and let the universe evolve with a fixed spatial geometry.

Like the de Sitter solution, however, these spacetimes are not very interesting. They are not quite static – they admit no timelike Killing vector – but they are *almost* static. Technically, time translation is a homothety: that is, there is a timelike vector ξ for which

$$\mathscr{L}_\xi g_{\mu\nu} = 2a g_{\mu\nu} \tag{3.90}$$

for some constant a. Okamura and Ishihara call these geometries 'static moduli solutions', and have shown that the set of such solutions is an attractor for nearby (small momentum) solutions [214, 215].

This lack of dynamics is clearly a consequence of our initial choice of vanishing momenta. Unfortunately, for $p \neq 0$ the Hamiltonian constraint (2.35) becomes much more difficult, and little is known beyond the formal existence and uniqueness theorems of Moncrief et al. [8, 202]. There appear to be two major problems:

1. For a surface of genus $g > 1$, there is no simple expression for the spatial metric analogous to that of equation (3.62) for the torus. Constant negative curvature spaces can be represented as regions of the hyperbolic plane $\mathbf{H}^2$ with appropriate identifications of boundary segments, and the problem of constructing such identifications from a set of moduli is fairly well understood. But in such a representation,

the moduli appear as boundary conditions rather than parameters in the metric, and the dynamics becomes much more difficult to describe.

2. The quadratic differentials p^{ij} are now position-dependent, and the Hamiltonian constraint is no longer solved by constant λ. While existence theorems are known, the actual construction of solutions becomes much more complicated, and very little can be said about explicit solutions.

A possible exception to these difficulties is the case of genus $g = 2$. Genus two surfaces are hyperelliptic; that is, they can be represented as branched double coverings of the sphere, with branch points determined by the moduli. This representation allows a reasonably simple description of the quadratic differentials, and it is plausible that the Hamiltonian constraint can be solved. For more general topologies, Puzio has suggested that the Gauss map, a harmonic map between a slice Σ of constant $Tr K$ and a fixed reference surface Σ_0, might serve as a useful tool, but this proposal has not yet been developed [223].

It may also be useful to look at decompositions of the metric that differ from those of chapter 2. In particular, any metric on a surface Σ is conformal to a *flat* metric with singularities and branch cuts. This flat representation has been studied extensively in string theory – it is the starting point for the light cone gauge [98] – and it is plausible that string techniques could be applied to the ADM constraints of (2+1)-dimensional gravity. It may also be useful to look at time-slicings that differ from the York gauge $Tr K = -T$. For systems of point particles in a topologically trivial spacetime, Bellini *et al.* and Welling have recently explored the use of conformally flat metrics on slices satisfying $Tr K = 0$ [33, 34, 279]. The solution of the vacuum field equations is then related to the classical Riemann–Hilbert problem, and a number of powerful mathematical techniques become available.

For the moment, however, solutions with the topology $[0, 1] \times \Sigma$ are well understood only for the torus and Klein bottle topologies. This is an unfortunate limitation, but we shall see in the next few chapters that even these simple examples can teach us a good deal about quantum gravity.

4

Geometric structures and Chern–Simons theory

In the two preceding chapters, we derived solutions of the vacuum field equations of (2+1)-dimensional gravity by using rather standard general relativistic methods. But as we have seen, the field equations in 2+1 dimensions actually imply that the spacetime metric is *flat* – the curvature tensor vanishes everywhere. This suggests that there might be a more directly geometric approach to the search for solutions.

At first sight, the requirement of flatness seems too strong: we usually think of the vanishing of the curvature tensor as implying that spacetime is simply Minkowski space. We have seen that this is not quite true, however. The torus universes of the last chapter, for example, are genuinely dynamical and have nontrivial – and inequivalent – global geometries. The situation is analogous to that of electromagnetism in a topologically nontrivial spacetime, where Aharanov–Bohm phases can be present even when the field strength $F_{\mu\nu}$ vanishes.

It is true, however, that *locally* we can always choose coordinates in which the metric is that of ordinary Minkowski space. That is, every point in a flat spacetime M is contained in a coordinate patch that is isometric to Minkowski space with the standard metric $\eta_{\mu\nu}$. The only place nontrivial geometry can arise is in the way these coordinate patches are glued together. This is precisely what we saw in chapter 3 for the spacetime surrounding a point source: locally, the geometry was flat, but a conical structure arose from the identification of the edges of a flat coordinate patch. The aim of this chapter is to generalize this construction to closed universes with more general topologies.

4.1 A static solution

Let us again consider the spacetimes with the topology $[0, 1] \times \Sigma$ discussed in chapters 2 and 3. We shall start in this section with one particularly simple family of spacetimes, and then generalize the results to incorporate the full set of solutions of the field equations.

We begin with a spacelike slice Σ, which we take to be a closed surface of genus $g \geq 2$. (We shall deal with the $g = 0$ and $g = 1$ cases later.) A classical result of two-dimensional geometry, the uniformization theorem, asserts that any such surface can be given a metric of constant negative curvature (see appendix A). A surface Σ with such a metric can be cut open along $2g$ curves to form a $4g$-sided polygon $P(\Sigma)$ whose sides are geodesics in the hyperbolic plane $\mathbf{H}^2$, that is, the plane $\mathbf{R}^2$ equipped with a constant negative curvature metric. Several useful representations of $\mathbf{H}^2$ exist, including the upper half-plane representation, characterized by the metric

$$ds^2 = \frac{k}{y^2}(dx^2 + dy^2), \qquad (y > 0), \tag{4.1}$$

and the Poincaré disk, characterized by the metric

$$ds^2 = \frac{4k\,dz\,d\bar{z}}{(1 - |z|^2)^2}, \qquad (|z| < 1). \tag{4.2}$$

The polygon $P(\Sigma)$ is most easily visualized in the Poincaré disk representation, in which geodesics are arcs of circles that meet the boundary $|z| = 1$ perpendicularly; for instance, figure 4.1 shows a polygon for a genus two surface. To recover the closed surface Σ from this polygon, we identify edges in pairs, in the pattern shown in figure A.8. Since the metric (4.2) is homogeneous, it will remain smooth under such identifications as long as the identified edges are geodesics with equal lengths, and Σ will thus inherit a smooth constant negative curvature metric.

To understand this 'gluing' operation, consider for a moment the simpler case of the flat torus. Let z be the standard coordinate on the complex plane $\mathbf{C}$, and let T denote the group of translations generated by $z \to z+1$ and $z \to z + \tau$, where τ is an arbitrary complex number with $Im\,\tau \neq 0$ (see figure 3.3). With a bit of thought, it should be clear that the quotient space $\mathbf{C}/T$ – that is, the space of points in $\mathbf{C}$ modulo the identifications $z \sim z + 1$ and $z \sim z + \tau$ – is a flat torus with modulus τ.

In the same way, the operation of gluing the sides of $P(\Sigma)$ to reconstruct Σ can be characterized by a discrete group Γ of isometries of the hyperbolic plane $\mathbf{H}^2$. A generator γ of Γ is a mapping that takes an edge E to the edge E' with which it is identified. Just as in the case of the torus, we can view Σ as a quotient space

$$\Sigma = \mathbf{H}^2/\Gamma \tag{4.3}$$

of the hyperbolic plane by a group of isometries. Not every group Γ can occur in this way, however. As discussed in appendix A, Γ must satisfy

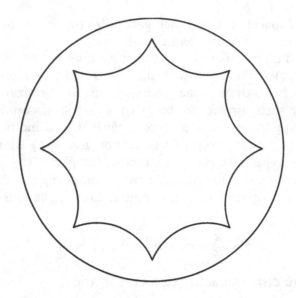

Fig. 4.1.   A genus two surface can be represented as an octagon on the Poincaré disk, with geodesic boundaries that must be identified pairwise (see also figure A.8).

two conditions in order to be a group of edge identifications of a polygon $P(\Sigma)$:

1. Γ must be isomorphic to the fundamental group $\pi_1(\Sigma)$; and

2. Γ must act freely and properly discontinuously on $\mathbf{H}^2$.

These conditions imply that Γ must be a discrete subgroup of the group $PSL(2, \mathbf{R})$ of isometries of $\mathbf{H}^2$ with no elements of finite order. Moreover, it is not hard to show that two such groups Γ_1 and Γ_2 determine identical surfaces if and only if they are conjugate, i.e., $\Gamma_2 = h\Gamma_1 h^{-1}$ for some fixed $h \in PSL(2, \mathbf{R})$.

A group satisfying these two conditions is called *Fuchsian*. While the complete characterization and classification of Fuchsian groups requires some fairly advanced mathematics, the subject has been studied extensively, and a large body of knowledge is available for our use [141].

Observe next that a surface of constant proper time in Minkowski space,

$$S_\tau = \{(t, x, y) : t^2 - x^2 - y^2 = \tau^2, \quad t > 0\}, \tag{4.4}$$

is isometric to the hyperbolic plane (4.2) with $k = \tau$. Let V^+ denote the interior of the forward light cone in Minkowski space. We can then perform the following construction:

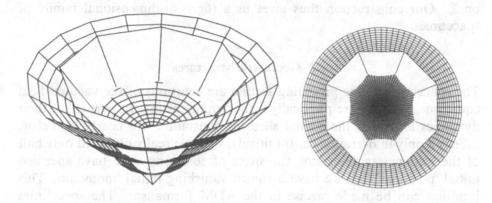

Fig. 4.2. A spacetime with the spatial topology of a genus two surface can be represented as an octagonal tube in Minkowski space, with edges identified in pairs. The left-hand illustration shows several slices of constant TrK. The right-hand illustration shows the intersection of one such slice with the octagon (compare figure 4.1).

1. foliate V^+ by slices of constant proper time;

2. draw an appropriately scaled copy of the polygon $P(\Sigma)$ on each slice S_τ, $a \leq \tau \leq b$, forming a figure with the topology $[0,1] \times P(\Sigma)$; and

3. glue the edges of this figure slice by slice to obtain a manifold with the topology $[0,1] \times \Sigma$.

Equivalently, we can start with the polygon $P(\Sigma)$ on a single slice, say $S_{\tau=1}$. Each edge of $P(\Sigma)$ determines a plane through the origin, and the set of such planes carves out a solid with the topology $[0,1] \times P(\Sigma)$ (see figure 4.2). The sides of this solid are identified pairwise by the action of the group Γ, and the manifold $[0,1] \times \Sigma$ is obtained by making these identifications. In terms of group theory, we have defined an action of the Fuchsian group Γ on V^+ by its action on each slice S_τ; our spacetime is the quotient space V^+/Γ. The resulting manifold has an initial singularity at the apex of the light cone, but away from this singularity it is flat, having inherited its metric from flat Minkowski space. We have thus built a solution of the $(2+1)$-dimensional empty space Einstein equations. In fact, we have found a whole family of solutions: each constant negative curvature metric on the surface Σ determines a polygon $P(\Sigma)$, and each such polygon determines a spacetime. If Σ is a genus g surface, the space of nonisometric polygons $P(\Sigma)$ is $(6g-6)$-dimensional. In fact, this space is precisely the moduli space $\mathcal{N}(\Sigma)$ introduced in chapter 3, since each inequivalent polygon determines a constant curvature $k = -1$ metric on Σ. Our construction thus gives us a $(6g - 6)$-dimensional family of spacetimes.

4.2 Geometric structures

The spacetimes of the preceding section are solutions of the vacuum field equations, but they are physically rather uninteresting. In particular, their dynamics is trivial; the spatial slices of constant τ are nearly isometric, differing only in overall scale. Intuitively, we have really obtained only half of the parameters describing the space of solutions – we have specified initial 'positions', but we have assumed vanishing initial 'momenta'. This intuition can be made precise in the ADM formalism. The spacetimes constructed in this manner are precisely the 'static moduli solutions' of section 4 of the preceding chapter, that is, the solutions for which $p^\alpha = 0$.

To generalize this result, it is useful to give a slightly different description of our construction. Let us return again to two dimensions. We can view the polygon $P(\Sigma)$ as a kind of coordinate patch on the surface Σ – that is, the interior of $P(\Sigma)$ is the diffeomorphic image of an open set $\mathcal{O}$ in Σ, and points in $\mathcal{O}$ can be identified with their images in $P(\Sigma)$. As in any manifold, coordinate patches are glued together by means of transition functions, which in this case describe the coordinate changes as we go from an edge E of $P(\Sigma)$ to the edge E' identified with E. $P(\Sigma)$ has a constant negative curvature metric, and the metric will remain smooth as long as the transition functions $\phi(E, E')$ are isometries. In fact, $P(\Sigma)$ has $4g$ edges identified in pairs, and the $2g$ transition functions $\phi(E_i, E_i')$ are precisely the generators of the Fuchsian group Γ introduced above.

What we have just described is known to mathematicians as a *geometric structure* [3, 58, 256]. In general, a manifold M is said to have a geometric structure (G, X) if M is locally modeled on X with transition functions in G, much as an ordinary n-dimensional manifold is modeled on $\mathbf{R}^n$. More precisely, let G be a Lie group that acts analytically on some manifold X (the 'model space'), and let M be another manifold of the same dimension as X. Then a (G, X) structure on M is given by a set of coordinate patches U_i on M with 'coordinates' $\phi_i : U_i \rightarrow X$ taking their values in X, and with transition functions $g_{ij} = \phi_i \circ \phi_j^{-1} | U_i \cap U_j$ in G. In particular, if we let X be the hyperbolic plane $\mathbf{H}^2$ and G be the isometry group $PSL(2, \mathbf{R})$, we obtain a *hyperbolic structure* on the surface Σ.

A fundamental ingredient in the description of a geometric structure is its *holonomy*, which can be thought of as measuring the failure of a single coordinate patch to extend all the way around a closed curve.* Let M be a (G, X) manifold containing a closed path γ. We can cover γ with

---

* The reader should be cautioned that this is not quite the same 'holonomy' that is found

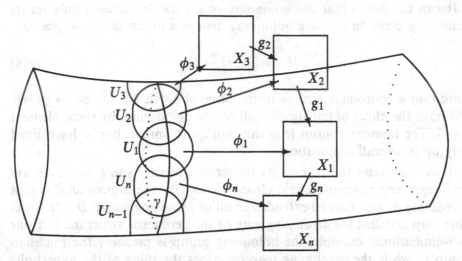

Fig. 4.3.   The curve γ is covered by patches $U_1, \ldots, U_n$, each isometric to a portion of the model space X. The composition of transition functions g_i around γ gives the holonomy.

coordinate charts

$$\phi_i : U_i \to X, \qquad i = 1, \ldots, n, \qquad (4.5)$$

with constant transition functions $g_i \in G$ between U_i and U_{i+1}, so

$$\phi_i | U_i \cap U_{i+1} = g_i \circ \phi_{i+1} | U_i \cap U_{i+1}$$
$$\phi_n | U_n \cap U_1 = g_n \circ \phi_1 | U_n \cap U_1 \qquad (4.6)$$

(see figure 4.3). Let us now try to analytically continue the coordinate ϕ_1 from U_1 to all of γ. We start with a coordinate transformation in U_2 that replaces ϕ_2 by $\phi_2' = g_1 \circ \phi_2$. By (4.6), ϕ_2' agrees with ϕ_1 on the overlap $U_1 \cap U_2$, and can thus be considered as an extension of ϕ_1 to the union $U_1 \cup U_2$. Continuing this process along γ, with $\phi_j' = g_1 \circ \ldots \circ g_{j-1} \circ \phi_j$, we will eventually reach the final coordinate patch U_n, which again overlaps U_1. If the new coordinate function $\phi_n' = g_1 \circ \ldots \circ g_{n-1} \circ \phi_n$ were equal to ϕ_1 on $U_n \cap U_1$, we would have succeeded in covering γ with a single patch. In general, however, we will find instead that $\phi_n' = H(\gamma) \circ \phi_1$, where

$$H(\gamma) = g_1 \circ \ldots \circ g_n. \qquad (4.7)$$

The holonomy $H(\gamma)$ thus measures the obstruction to such a covering.

---

in the theory of connections on fiber bundles, although we shall see below that the two notions are closely related.

It can be shown that the holonomy of a curve γ depends only on its homotopy class. In fact, the holonomy defines a group homomorphism

$$H : \pi_1(M, *) \to G. \tag{4.8}$$

Note that a coordinate change in the original patch U_1, say $\phi_1 \to \phi_1 \circ h$, will have the effect of conjugating all of the holonomies by some element $h \in G$. The homomorphism H is thus not quite unique, but is determined only up to overall conjugation.

If we now pass from M to its universal covering space $\widetilde{M}$, there are no longer any noncontractible closed paths. By the argument above, a coordinate ϕ_1 can thus be extended to all of $\widetilde{M}$, giving a map $D : \widetilde{M} \to X$. This map is called the *developing map* of the geometric structure. For our two-dimensional example, the holonomy group is precisely the Fuchsian group Γ, while the developing map describes the tiling of the hyperbolic plane by copies of the polygon $P(\Sigma)$.

As a concrete example of this construction, consider the conical spacetime of a static point particle discussed in chapter 3. At a fixed time, space is described by a wedge W (figure 3.1), with edges identified by a rotation $R(\beta)$; the whole spacetime is $\mathbf{R} \times W$, with the same identification at all times. The fundamental group of such a spacetime has a single generator γ, describing a loop around the apex of the cone. The corresponding holonomy is $H(\gamma) = R(\beta)$, and the developing map is the map that 'un-wraps' the cone into an infinite strip. Similarly, for a spinning particle, the holonomy is a Poincaré transformation that includes a rotation and a time translation.

We can now describe the three-dimensional cosmologies of the preceding section in the same language. We start with the region $[0, 1] \times P(\Sigma)$ of figure 4.2, and identify sides $[0, 1] \times E_i$ and $[0, 1] \times E_i'$. Once again, $[0, 1] \times P(\Sigma)$ may be viewed as a coordinate patch with transition functions connecting the boundaries. Moreover, these transition functions may now be treated as isometries of the full $(2+1)$-dimensional Minkowski metric, thus describing a geometric structure.

But while these transition functions are isometries of the Minkowski metric, they lie in a rather restricted group of such isometries. The full group of isometries of the Minkowski metric is the Poincaré group $ISO(2, 1)$. The identifications of section 1, on the other hand, are isometries of the two-dimensional metric (4.2), and thus lie in the subgroup $PSL(2, \mathbf{R}) \subset ISO(2, 1)$. The generalization we need is now clear: we must allow transition functions that live in the full Poincaré group. In other words, we must look for *Lorentzian structures*, or $(ISO(2, 1), \mathbf{R}^{2+1})$ structures, on $[0, 1] \times \Sigma$.

4.3 The space of Lorentzian structures

To understand the classification of Lorentzian structures, we must first deal with a technical problem arising from the fact that the space $\mathbf{R} \times \Sigma$ is not compact. If M is a three-manifold with a flat Lorentzian metric, we can always find a 'new' noncompact solution to the field equations by simply cutting out a chunk of M. For instance, if we choose some spacelike slice S in M and remove the entire future of S, we will be left with a manifold with the same topology and the same geometric structure. To avoid this rather trivial overcounting of solutions, we need some notion of a 'maximal' spacetime. The appropriate concept turns out to be that of a domain of dependence, essentially a spacetime that includes the entire past and future of some spacelike slice (see appendix B for a more precise definition).

Mess has shown that every maximal flat Lorentzian spacetime with the topology $[0, 1] \times \Sigma$ is uniquely determined by its $ISO(2, 1)$ holonomy group [195]. Moreover, any homomorphism from $\pi_1(\Sigma)$ to $ISO(2, 1)$ gives an admissible holonomy group, provided only that its $SO(2, 1)$ projection is Fuchsian. In other words, no new restrictions are needed on the translations in the Poincaré group; as long as the 'positions' generate a Fuchsian subgroup of $SO(2, 1)$, the 'momenta' are arbitrary.

As always, two holonomy groups give different geometric structures unless they are conjugate. The solutions of the field equations are thus parametrized by the space of group homomorphisms

$$\widetilde{\mathscr{M}} = Hom_0(\pi_1(\Sigma), ISO(2, 1))/ \sim, \tag{4.9}$$

where

$$\rho_1 \sim \rho_2 \quad \text{if} \quad \rho_2 = h \cdot \rho_1 \cdot h^{-1}, \qquad h \in ISO(2, 1). \tag{4.10}$$

The subscript 0 in equation (4.9) means that we restrict ourselves to homomorphisms whose $SO(2, 1)$ projections are Fuchsian; Goldman has shown that this subspace is a connected topological component of the space of all homomorphisms [130].

One more subtlety must be taken into account if we want a complete description of the space of solutions. Diffeomorphic spacetimes should not be counted as distinct solutions, and while the space $\widetilde{\mathscr{M}}$ is not affected by diffeomorphisms that can be smoothly deformed to the identity, it *is* acted on by 'large' diffeomorphisms. To understand this phenomenon, note first that the large diffeomorphisms of $[0, 1] \times \Sigma$ can be characterized completely by their actions on Σ. The group of large diffeomorphisms of a surface Σ, also known as the mapping class group, is discussed in appendix A. It is generated by Dehn twists around closed curves in Σ, that is, diffeomorphisms obtained by cutting Σ open along a curve γ, twisting one end by 2π, and regluing the ends. (See figure A.1 for an illustration of a Dehn twist of a torus.) It is fairly easy to see that such Dehn twists will mix up the generators of the fundamental group $\pi_1(\Sigma)$; in fact, it may be shown that they generate all of the outer automorphisms of $\pi_1(\Sigma)$ [37].

This means that if $\rho_1 \in \widetilde{\mathscr{M}}$ and $\rho_2 \in \widetilde{\mathscr{M}}$ are two homomorphisms that differ only by an automorphism of $\pi_1(\Sigma)$, they will give the same flat metric on $[0,1] \times \Sigma$. Equivalently, the geometry of spacetime should depend only on the fundamental group $\pi_1(\Sigma)$ as an abstract group, and not on any particular choice of generators. The true space of flat spatially closed $(2+1)$-dimensional spacetimes is thus parametrized by

$$\mathscr{M} = \widetilde{\mathscr{M}}/Out(\pi_1(\Sigma)). \qquad (4.11)$$

This characterization of $\mathscr{M}$ makes it simple to determine the number of empty space solutions to the Einstein equations. A heuristic counting argument proceeds as follows. The fundamental group of a genus g surface Σ has $2g$ generators and one relation. Each generator is mapped into the six-dimensional space $ISO(2,1)$, giving a total of $6(2g-1)$ parameters. Conjugate groups must still be identified, however, so six of these parameters are redundant. We thus expect there to be a $(12g-12)$-dimensional space of solutions. This result confirms the ADM computations of chapter 2, in which we found that solutions were labeled by $6g-6$ positions and $6g-6$ momenta. The counting is slightly more complicated when Σ is a sphere or a torus, but it is not hard to check that the number of solutions again agrees with the ADM results.

We can establish an even closer connection with the ADM method if we note that the group $ISO(2,1)$ is geometrically the cotangent bundle of $SO(2,1)$; that is, translations in $ISO(2,1)$ can be viewed as cotangent vectors to Lorentz transformations. Indeed, if $t \mapsto \Lambda(t)$ is a curve in the group manifold $ISO(2,1)$, a cotangent vector at $\Lambda(0)$ is a vector of the form

$$v = \frac{d\Lambda(t)}{dt}\Lambda^{-1}(t)\Big|_{t=0}. \qquad (4.12)$$

Given two curves $\Lambda_1(t)$ and $\Lambda_2(t)$ through a common point in the group manifold of $SO(2,1)$, the 'product cotangent vector' is thus

$$v = \left(\frac{d}{dt}(\Lambda_1\Lambda_2)(t)\right)(\Lambda_1\Lambda_2)^{-1}(t)\Big|_{t=0} = v_1 + \Lambda_1(0)v_2\Lambda_1^{-1}(0), \qquad (4.13)$$

giving precisely the right semidirect product structure for $ISO(2,1)$. This result implies, in turn, that $\mathscr{M}$ is itself a cotangent bundle, $\mathscr{M} = T^*\mathscr{N}$, with a base space

$$\mathscr{N} = \widetilde{\mathscr{N}}/Out(\pi_1(\Sigma)), \qquad (4.14)$$

with

$$\widetilde{\mathcal{N}} = Hom_0(\pi_1(\Sigma), SO(2,1))/\sim,$$

$$\rho_1 \sim \rho_2 \quad \text{if} \quad \rho_2 = h \cdot \rho_1 \cdot h^{-1}, \qquad h \in SO(2,1). \tag{4.15}$$

This characterization will be crucial when we consider the quantization of (2+1)-dimensional gravity in the connection representation.

Observe that $\mathcal{N}$ is once again the ordinary moduli space of Σ, that is, the space of constant negative curvature metrics, or hyperbolic structures, on Σ. This again confirms the structure we found for the reduced phase space of chapter 2, in which the momenta p^α were cotangents to the moduli m_α.

4.4 Adding a cosmological constant

A similar construction exists when the cosmological constant is nonzero, but some new subtleties appear. Let us first take Λ to be negative, say $\Lambda = -1/\ell^2$. In this case, the model space analogous to Minkowski space is (2+1)-dimensional anti-de Sitter space (adS), or, strictly speaking, the universal covering space $\widetilde{adS}$. This space may be represented concretely by starting with the flat four-dimensional space $\mathbf{R}^{2,2}$, with coordinates (X_1, X_2, T_1, T_2) and metric

$$dS^2 = dX_1^2 + dX_2^2 - dT_1^2 - dT_2^2, \tag{4.16}$$

and restricting to the submanifold

$$X_1^2 + X_2^2 - T_1^2 - T_2^2 = -\ell^2. \tag{4.17}$$

We can write the coordinates of $\mathbf{R}^{2,2}$ in matrix form,

$$\mathbf{X} = \frac{1}{\ell}\begin{pmatrix} X_1 + T_1 & X_2 + T_2 \\ X_2 - T_2 & -X_1 + T_1 \end{pmatrix}, \tag{4.18}$$

and anti-de Sitter space is then determined by the condition

$$\det \mathbf{X} = 1, \tag{4.19}$$

i.e., $\mathbf{X} \in SL(2,\mathbf{R})$. In this representation, it is easy to check that the isometry group of $\widetilde{adS}$ is $SL(2,\mathbf{R}) \times SL(2,\mathbf{R})/\mathbf{Z}_2$, where the two factors of $SL(2,\mathbf{R})$ act by left and right multiplication, $\mathbf{X} \to R^+ \mathbf{X} R^-$, with $(R^+, R^-) \sim (-R^+, -R^-)$.

An anti-de Sitter structure is thus obtained by gluing together patches of $\widetilde{adS}$ with transition functions in $SL(2,\mathbf{R}) \times SL(2,\mathbf{R})/\mathbf{Z}_2$. As in the preceding section, the space of holonomies on a manifold $[0, 1] \times \Sigma$ with Σ spacelike is

$$\widetilde{\mathcal{M}} = Hom_0(\pi_1(\Sigma), SL(2, \mathbf{R}) \times SL(2, \mathbf{R})/\mathbf{Z}_2)/ \sim, \tag{4.20}$$

where

$$\rho_1 \sim \rho_2 \quad \text{if} \quad \rho_2 = h \cdot \rho_1 \cdot h^{-1}, \qquad h \in SL(2, \mathbf{R}) \times SL(2, \mathbf{R})/\mathbf{Z}_2. \tag{4.21}$$

The subscript 0 in (4.20) again means that we must restrict ourselves to homomorphisms whose projections on each of the $SL(2, \mathbf{R})$ factors are Fuchsian.

As in the case of a vanishing cosmological constant, it may be shown that a holonomy $\rho \in \widetilde{\mathcal{M}}$ determines a unique maximal spacetime, and that such spacetimes are parametrized by $\widetilde{\mathcal{M}}$ modulo the action of the mapping class group [195]. In contrast to the $\Lambda = 0$ case, however, the space (4.20) is not a cotangent bundle. Rather, $\widetilde{\mathcal{M}}$ is, roughly speaking, a product of two copies of the ordinary moduli space (4.15) of Σ, each with its own symplectic structure.

For the case of a positive cosmological constant, the relevant model space is de Sitter space, which can be obtained from $\mathbf{R}^{1,3}$, with coordinates (T, X, Y, Z) and metric

$$dS^2 = -dT^2 + dX^2 + dY^2 + dZ^2, \tag{4.22}$$

by restricting to the submanifold

$$-T^2 + X^2 + Y^2 + Z^2 = -\frac{1}{\Lambda}. \tag{4.23}$$

It is evident from this construction that (2+1)-dimensional de Sitter space has isometry group $SO(3, 1)$. As in the previous cases, solutions of the field equations with $\Lambda > 0$ may be described by de Sitter geometric structures, that is, by piecing together patches of de Sitter space with transition functions in $SO(3, 1)$.

For the case of de Sitter structures, however, Mess has shown that the holonomies do *not* determine the geometry [195]. Rather, an infinite discrete set of spacetimes have the same holonomy. We shall see a specific example of this phenomenon at the end of the next section. This ambiguity may indicate the presence of a new discrete quantum number in quantum gravity. Alternatively, as Witten has argued, we may choose to take the holonomies as the fundamental observables; in that case, spacetimes that differ classically would be quantum mechanically indistinguishable [290].

4.5 Closed universes as quotient spaces

In sections 1 and 2, we encountered two equivalent descriptions of a two-dimensional surface, as a quotient space $\mathbf{H}^2/\Gamma$ and as a geometric space modeled on $\mathbf{H}^2$ with holonomy group Γ. We succeeded in generalizing the latter representation to three dimensions. It is natural to ask whether the quotient space representation can be extended as well.

In general, it cannot. Consider, for example, the conical spacetime of a point particle with an irrational deficit angle β. Rotations by β are not periodic, and enough rotations will bring an initial point x_0 arbitrarily close to any other point x at the same radius. Consequently, the group generated by the rotation $R(\beta)$ does not act properly discontinuously – it does not 'separate points' – and the quotient space $\mathbf{R}^2/\langle R(\beta)\rangle$ is not well-behaved.[†]

For cosmological solutions with the topology $[0,1] \times \Sigma$, on the other hand, the situation is more favorable. Mess has shown that if Σ is spacelike, any such spacetime can be written as a quotient space N/Γ, where Γ is the holonomy group and (for genus greater than one) N is a region in the interior of the light cone of Minkowski space [195]. In principle, this makes it possible to construct the spacetime $[0,1] \times \Sigma$ explicitly: one need merely find a set of coordinates upon which the group Γ acts nicely and form the quotient space by identifying appropriate coordinate values.

In practice, this task is already almost unmanageable in two dimensions. Even for genus two, such coordinate identifications are extraordinarily difficult to describe explicitly. For genus one, however – that is, for spacetimes with the spatial topology of a torus – the quotient space construction allows a complete, simple, and explicit description of all flat Lorentzian metrics with spacelike hypersurfaces T^2. We now turn to this simple case.

We must first determine the possible holonomy groups of the spacetime $M = [0,1] \times T^2$. The fundamental group of the torus, and thus of M, is the abelian group $\mathbf{Z} \oplus \mathbf{Z}$, with one generator for each of the two independent circumferences of T^2. The holonomy group must therefore be generated by two commuting Poincaré transformations, say (Λ_1, a_1) and (Λ_2, a_2).

We begin by analyzing the $SO(2,1)$ components Λ_1 and Λ_2. Any Lorentz transformation in 2+1 dimensions fixes a vector n, and for Λ_1 and Λ_2 to commute, they must fix the same vector. This vector may be spacelike,

---

[†] For the cone, there is a way out: if we remove the line $r=0$ from Minkowski space, and then form the universal covering space (by allowing the polar angle ϕ to range from $-\infty$ to ∞), the conical spacetime can be expressed as the quotient of this covering space by $R(\beta)$. Similar constructions have been found in other specific instances, but a systematic generalization is not known.

null, or timelike, and the space of holonomies correspondingly splits into three components.

(A similar splitting occurs for spaces of higher genus as well. For $g > 1$, the components are genuinely topologically disconnected, and only one component corresponds to a Lorentzian structure on $[0, 1] \times \Sigma$ [130]. For the torus, on the other hand, the components are not completely disconnected, and the space of holonomies is a complicated, non-Hausdorff space. For a careful description of this space, see reference [181].)

If n is timelike or null, it may be shown that the toroidal slices T^2 are not spacelike: roughly speaking, a T^2 slice must contain n. In this case, the quotient construction fails. If n is timelike, for example, Λ_1 and Λ_2 are conjugate to rotations, and the problem is essentially the same as that of constructing a conical geometry as a quotient space. Nonstandard quotient constructions may still be possible, as described in the footnote on page 71; for the toroidal universe with timelike or null n, this possibility is explored in references [108, 181]. For most 'physical' applications, however, we are interested in solutions like those that arise in the ADM description, for which the T^2 slices are spacelike.

For such solutions, two possibilities remain: either Λ_1 and Λ_2 fix a spacelike vector, say $(0, 0, 1)$, or Λ_1 and Λ_2 both vanish. Let us start with the latter, simpler possibility. The holonomy group is now generated by a pair of translations a_1 and a_2, which we take to be spacelike in order to ensure that the T^2 slices are spacelike. We may then choose coordinates such that $a_1 = (0, a, 0)$ and $a_2 = (0, a\tau_1, a\tau_2)$. A fundamental region for the action of the holonomy group on a spatial slice of constant t is now a parallelogram with vertices at $(0, 0)$, $(a, 0)$, $(a\tau_1, a\tau_2)$, and $(a(\tau_1 + 1), a\tau_2)$. The spatial geometry on each slice is that of a torus with modulus $\tau = \tau_1 + i\tau_2$ and area $a\tau_2$, and the resulting spacetime is precisely the static torus geometry (3.83).

If, on the other hand, Λ_1 and Λ_2 stabilize a spacelike vector, they must both be boosts. We can then use our remaining freedom of overall conjugation to put the two generators of the holonomy group in the form

$$H(\gamma_1) : (t, x, y) \to (t \cosh \lambda + x \sinh \lambda, x \cosh \lambda + t \sinh \lambda, y + a)$$
$$H(\gamma_2) : (t, x, y) \to (t \cosh \mu + x \sinh \mu, x \cosh \mu + t \sinh \mu, y + b).$$
$$(4.24)$$

This action simplifies if we define new coordinates

$$t = \frac{1}{T} \cosh u, \qquad x = \frac{1}{T} \sinh u, \qquad (4.25)$$

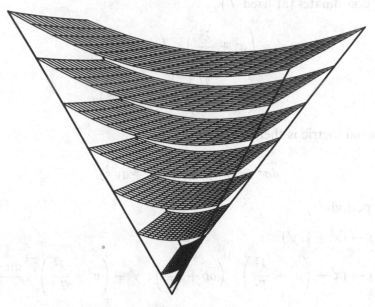

Fig. 4.4.   A flat torus universe is foliated by surfaces of constant TrK. Each slice is a parallelogram (with opposite edges identified), but the area and modulus vary from slice to slice. Note that at the initial singularity, the parallelogram collapses to a line.

in which the Minkowski metric becomes

$$ds^2 = -\frac{1}{T^4}dT^2 + \frac{1}{T^2}du^2 + dy^2. \qquad (4.26)$$

The coordinate T has been chosen so that surfaces of constant T have extrinsic curvature $TrK = -T$, in order to allow comparison with the results of chapter 3.

In the new coordinates, the transformations (4.24) reduce to

$$H(\gamma_1) : (T, u, y) \to (T, u + \lambda, y + a)$$
$$H(\gamma_2) : (T, u, y) \to (T, u + \mu, y + b). \qquad (4.27)$$

On a constant T surface, a fundamental region for the action of the holonomy group $\langle H(\gamma_1), H(\gamma_2) \rangle$ is thus simply the torus $(u, y) \sim (u + \lambda, y + a) \sim (u + \mu, y + b)$. A typical three-manifold obtained by such identifications is illustrated in figure 4.4.

To put this metric in a more standard form, we can transform to new spatial coordinates (at fixed T)

$$x' = \left(a^2 + \frac{\lambda^2}{T^2}\right)^{-1}\left(ay + \frac{\lambda}{T^2}u\right)$$

$$y' = \left(a^2 + \frac{\lambda^2}{T^2}\right)^{-1}\left(\frac{\lambda y - au}{T}\right). \tag{4.28}$$

The spatial metric is then

$$d\sigma^2 = (a^2 + \frac{\lambda^2}{T^2})(dx'^2 + dy'^2), \tag{4.29}$$

with a periodicity

$$(x', y') \rightarrow (x' + 1, y')$$

$$(x', y') \rightarrow (x' + \left(a^2 + \frac{\lambda^2}{T^2}\right)^{-1}\left(ab + \frac{\lambda\mu}{T^2}\right), \, y' + \left(a^2 + \frac{\lambda^2}{T^2}\right)^{-1}\frac{a\mu - \lambda b}{T}).$$

$$\tag{4.30}$$

From the definition of the modulus τ, $d\sigma^2$ is therefore the metric of a torus with

$$\tau = \tau_1 + i\tau_2 \tag{4.31}$$

$$= \left(a^2 + \frac{\lambda^2}{T^2}\right)^{-1}\left[\left(ab + \frac{\lambda\mu}{T^2}\right) + i\left(\frac{a\mu - \lambda b}{T}\right)\right] = \left(a + \frac{i\lambda}{T}\right)^{-1}\left(b + \frac{i\mu}{T}\right).$$

The corresponding momentum conjugate to τ is easily found to be

$$p = -iT\left(a - \frac{i\lambda}{T}\right)^2. \tag{4.32}$$

In contrast to the case of purely translational holonomies, such a metric is clearly dynamical: the shape of a toroidal cross-section changes with time. Indeed, the solutions we have constructed are precisely the four-parameter family of solutions (3.80) that we already found in chapter 3, with $\Lambda = 0$.

Let us next attempt to generalize this quotient construction to the case of a nonvanishing cosmological constant. For $\Lambda = -1/\ell^2 < 0$, maximal spacetimes are again determined by their (anti-de Sitter) holonomies, and most of the preceding discussion can be carried through with only small changes. For the $[0, 1] \times T^2$ topology, for example, the holonomies now fall into six sectors, since the isometry in each $SL(2, \mathbf{R})$ factor can fix a timelike, null, or spacelike vector. Spacelike T^2 slices occur in the 'spacelike–spacelike' sector, in which the two commuting holonomies

$(R_1^+, R_1^-) \in SL(2,\mathbf{R}) \times SL(2,\mathbf{R})$ and $(R_2^+, R_2^-) \in SL(2,\mathbf{R}) \times SL(2,\mathbf{R})$ may be conjugated to the form

$$R_a^\pm = \begin{pmatrix} e^{r_a^\pm/2} & 0 \\ 0 & e^{-r_a^\pm/2} \end{pmatrix}. \tag{4.33}$$

These holonomies act on the coordinates X_1, X_2, T_1, and T_2 of equation (4.16), and the geometry of the resulting quotient space can be translated back into the form (3.80) of chapter 3 through the coordinate transformation

$$X_1 = Fe^{-t/2} \sinh(\lambda x + \mu y)$$
$$T_1 = Fe^{-t/2} \cosh(\lambda x + \mu y)$$
$$X_2 = F\ell e^{t/2} \sinh\left(\frac{ax + by}{\ell}\right)$$
$$T_2 = F\ell e^{t/2} \cosh\left(\frac{ax + by}{\ell}\right), \tag{4.34}$$

with

$$a = \frac{\ell}{2}(r_2^+ - r_2^-), \qquad b = \frac{\ell}{2}(r_1^+ - r_1^-)$$
$$\lambda = \frac{1}{2}(r_2^+ + r_2^-), \qquad \mu = \frac{1}{2}(r_1^+ + r_1^-). \tag{4.35}$$

A straightforward computation then shows that the isometries generated by the holonomies (4.33) reduce to the identifications $x \sim x + 1$ and $y \sim y + 1$.

A slice of this spacetime at fixed York time has a two-geometry that can be easily computed from equation (3.81) of chapter 3. The results are most elegantly expressed in terms of a new time coordinate $\bar{t}$, defined by the condition

$$T = -\frac{2}{\ell} \cot \frac{2\bar{t}}{\ell}, \tag{4.36}$$

or in terms of the coordinate t of equation (3.68),

$$e^t = \frac{1}{\ell} \cot \frac{\bar{t}}{\ell}. \tag{4.37}$$

One finds that

$$\tau = \left(r_2^- e^{i\bar{t}/\ell} + r_2^+ e^{-i\bar{t}/\ell}\right)^{-1} \left(r_1^- e^{i\bar{t}/\ell} + r_1^+ e^{-i\bar{t}/\ell}\right)$$
$$p = \frac{i\ell}{2\sin(2\bar{t}/\ell)} \left(r_2^+ e^{i\bar{t}/\ell} + r_2^- e^{-i\bar{t}/\ell}\right)^2$$
$$H_{red} = \frac{\ell^2}{4} \sin \frac{2\bar{t}}{\ell} (r_1^- r_2^+ - r_1^+ r_2^-). \tag{4.38}$$

By (3.82), the symplectic structure is now

$$\delta p_1 \wedge \delta \tau_1 + \delta p_2 \wedge \delta \tau_2 = \ell \left(\delta r_1^+ \wedge \delta r_2^+ - \delta r_1^- \wedge \delta r_2^- \right),$$

(4.39)

so (r_1^+, r_2^+) and (r_1^-, r_2^-) are the two conjugate pairs. This is a particular case of a more general phenomenon: when $\Lambda < 0$, the two copies of $SL(2, \mathbf{R})$ that define the geometric structure have independent symplectic structures. Note that the Hamiltonian in (4.38) generates translations in T; the corresponding generator of translations in $\bar{t}$ is

$$H'_{\bar{t}} = \frac{dT}{d\bar{t}} H_{red} = \csc \frac{2\bar{t}}{\ell} \left(r_1^- r_2^+ - r_1^+ r_2^- \right) = \frac{2}{\ell} \csc \frac{2\bar{t}}{\ell} \tau_2 \left(p_1{}^2 + p_2{}^2 \right)^{1/2}.$$

(4.40)

A similar quotient construction exists for the (2+1)-dimensional black hole [71]. Indeed, the identifications (3.61) of chapter 3 are an isometry of anti-de Sitter space, and can be represented by the $SL(2, \mathbf{R}) \times SL(2, \mathbf{R})$ holonomy (R^+, R^-) with

$$R^+ = \begin{pmatrix} e^{\pi(r_+ - r_-)/\ell} & 0 \\ 0 & e^{-\pi(r_+ - r_-)/\ell} \end{pmatrix},$$

$$R^- = \begin{pmatrix} e^{\pi(r_+ + r_-)/\ell} & 0 \\ 0 & e^{-\pi(r_+ + r_-)/\ell} \end{pmatrix},$$

(4.41)

where $r_\pm$ are the inner and outer horizons (3.46). This quotient construction will be discussed further in chapter 12.

For $\Lambda > 0$, the situation is rather different. The relevant simply connected model space is now de Sitter space, and the holonomies lie in the isometry group $SO(3, 1)$. In this case, it is *not* true that the holonomies uniquely determine the geometric structure; as discussed in the preceding section, a given holonomy typically corresponds to an infinite number of distinct spacetimes.

For the torus topology, for example, a pair of commuting holonomies may be put in the form

$$R_{1,2} = \begin{pmatrix} \cosh \alpha_{1,2} & \sinh \alpha_{1,2} & 0 & 0 \\ \sinh \alpha_{1,2} & \cosh \alpha_{1,2} & 0 & 0 \\ 0 & 0 & \cos u_{1,2} & -\sin u_{1,2} \\ 0 & 0 & \sin u_{1,2} & \cos u_{1,2} \end{pmatrix}.$$

(4.42)

These act on the coordinates (T, X, Y, Z) of equation (4.22) by ordinary matrix multiplication. The coordinate transformation analogous to (4.34) is now

$$T = Fe^{-t/2}\cosh(\lambda x + \mu y)$$
$$X = Fe^{-t/2}\sinh(\lambda x + \mu y)$$
$$Y = F\Lambda^{-1/2}e^{t/2}\cos\Lambda^{1/2}(ax + by)$$
$$Z = F\Lambda^{-1/2}e^{t/2}\sin\Lambda^{1/2}(ax + by), \qquad (4.43)$$

where

$$a = \Lambda^{-1/2}u_1, \qquad b = \Lambda^{-1/2}u_2,$$
$$\lambda = \alpha_1, \qquad \mu = \alpha_2. \qquad (4.44)$$

The modulus and momentum can again be computed from (3.81):

$$\tau = \left(u_1 + i\Lambda^{1/2}e^{-t}\alpha_1\right)^{-1}\left(u_2 + i\Lambda^{1/2}e^{-t}\alpha_2\right)$$
$$p = -i\Lambda^{-1}e^t\left(u_1 - i\Lambda^{1/2}e^{-t}\alpha_1\right)^2. \qquad (4.45)$$

So far, these expressions look quite similar to those for a negative cosmological constant. Note, though, that the holonomies (4.42) are *periodic* in the parameters u_1 and u_2, while the modulus and momentum clearly are not. Thus the holonomies do not uniquely determine the metric.

This loss of periodicity was secretly introduced in the coordinate transformation (4.43), which is not periodic, and one might worry that this has invalidated the derivation. But it may be checked explicitly that if the parameters (4.44) are inserted in the metric (3.80) of chapter 3, they yield a smooth solution of the Einstein field equations with the topology $[0, 1] \times T^2$ and with holonomies (4.42). We have thus found an infinite family of metrics, characterized by parameters $(\mu, \lambda, a + 2\pi\Lambda^{-1/2}n_1, b + 2\pi\Lambda^{-1/2}n_2)$, that have identical $SO(3, 1)$ holonomies [106].

4.6 Fiber bundles and flat connections

We saw in chapter 2 that $(2+1)$-dimensional gravity with $\Lambda = 0$ can be rewritten as a Chern–Simons theory for a vector potential with gauge group $ISO(2, 1)$. The classical solutions of a Chern–Simons theory are gauge fields with vanishing field strength, that is, flat $ISO(2, 1)$ connections. It is thus interesting to see whether we can reexpress the geometric descriptions of the last two sections in the language of connections on fiber bundles.[‡]

[‡] I will assume for this section that the reader is familiar with the basic features of connections on fiber bundles, as summarized briefly in appendix C. For a more complete introduction, the books [159] and [204] give a good 'physicist's description'.

Recall that a (G, X) structure on M is determined by a set of coordinate patches U_i homeomorphic to X, with transition functions $\phi_i \circ \phi_j^{-1}$ in G. Let $\mathscr{G}$ be the Lie algebra of G. $\mathscr{G}$ is a vector space, and we can construct a flat vector bundle with base space M and fiber $\mathscr{G}$ as follows:

1. For each patch U_i, form the product $U_i \times \mathscr{G}$;

2. On each overlap $U_i \cap U_j$, take as a transition function for the fibers the adjoint action of $\phi_i \circ \phi_j^{-1} \in G$ acting on $\mathscr{G}$.

In the special case that M can be written as a quotient space $M = X/\Gamma$, this is equivalent to forming the bundle

$$E = (X \times \mathscr{G})/\Gamma, \tag{4.46}$$

where the quotient is by the simultaneous action of Γ on X and $\mathscr{G}$,

$$(x, v) \sim (gx, g^{-1}vg), \qquad x \in X, \; v \in \mathscr{G}. \tag{4.47}$$

The bundle E constructed in this manner has a natural flat $\mathscr{G}$-valued connection, which can be defined as follows. On some initial patch U_1, set $A_\mu = 0$. On an adjacent patch U_2, it follows from (4.6) that

$$A_\mu | U_2 = g_1^{-1} \partial_\mu g_1, \tag{4.48}$$

where $g_1 = \phi_1 \circ \phi_2^{-1}$. Continuing this process, we can determine the connection throughout M.

Now let γ be a closed curve starting and ending in the patch U_1. To calculate the holonomy of the connection A along this curve, we must parallel transport a vector around γ. From (4.48) and equation (C.22) of appendix C, parallel transport from U_1 to U_2 is determined by the equation

$$\frac{dv}{dt} + g_1^{-1}\frac{dg_1}{dt}v = 0, \tag{4.49}$$

which implies that

$$v|U_2 = g_1^{-1}v|U_1. \tag{4.50}$$

Continuing this process around a chain of patches as in figure 4.3, we find that

$$v|U_1 \rightarrow g_n^{-1} \ldots g_1^{-1} v|U_1. \tag{4.51}$$

The holonomy of the connection A is thus

$$H(\gamma) = g_n^{-1} \ldots g_1^{-1}, \tag{4.52}$$

which is essentially the same as the holonomy of the geometric structure defined in section 2. (The expression (4.52) is actually the inverse of the holonomy of the geometric structure found previously, but the difference is merely one of convention; traversing γ in the opposite direction would invert the holonomy of the connection.)

We can now specialize to the case $G = ISO(2, 1)$. We saw above that a solution of the Einstein equations in 2+1 dimensions is given by a Lorentzian structure on a spacetime manifold M. It is now apparent that we can equally well specify a flat $ISO(2, 1)$ connection on M. For the topologies we are considering, in which the holonomy determines the geometric structure, the two approaches are completely equivalent, thus confirming the equivalence of the metric and Chern–Simons formulations of the field equations.

This equivalence can be made rather concrete: given a flat $ISO(2, 1)$ connection (e, ω), we can write down an explicit differential equation for the developing map D of page 66 [3]. Let U be an open set in a spacetime M, and consider a function q from U to Minkowski space $\mathbf{R}^{2,1}$ that satisfies

$$dq^a + \epsilon^{abc}\omega_b q_c + e^a = 0. \tag{4.53}$$

It is easy to check that the integrability conditions for this equation are the first-order field equations of chapter 2,

$$T^a = de^a + \epsilon^{abc}\omega_b e_c = 0$$
$$R^a = d\omega^a + \frac{1}{2}\epsilon^{abc}\omega_b\omega_c = 0. \tag{4.54}$$

If we choose a gauge such that $\omega^a = 0$ in U – such a choice is always possible for a flat connection – then (4.53) implies that

$$g_{\mu\nu} = e_\mu{}^a e_{\nu a} = \eta_{ab}\partial_\mu q^a \partial_\nu q^b, \tag{4.55}$$

so the q^a can be viewed as local coordinates in a patch of Minkowski space. The conditions (4.54) guarantee only local integrability, but if we lift e and ω from M to its universal covering space $\widetilde{M}$, there are no obstructions to globally integrating (4.53). We can thus treat q as a map $q : \widetilde{M} \to \mathbf{R}^{2,1}$.

To understand the global properties of this map, we must investigate its behavior under gauge transformations. The infinitesimal $ISO(2, 1)$ transformations of (e, ω) were given in equations (2.66)–(2.67); for a finite transformation $(\Lambda, b) \in ISO(2, 1)$, the integrated version is

$$e^a \to \Lambda^a{}_b e^b - db^a + (d\Lambda^a{}_c)\Lambda_b{}^c b^b + \epsilon^{abc}\Lambda_{cd}b_b\omega^d$$
$$\omega^a \to \Lambda^a{}_b\omega^b + \frac{1}{2}\epsilon^{abc}(d\Lambda_b{}^d)\Lambda_{cd}. \tag{4.56}$$

It is straightforward to check that the transformation of q required to leave (4.53) invariant is then

$$q^a \to \Lambda^a{}_b q^b + b^a, \tag{4.57}$$

which may be recognized as the standard $ISO(2,1)$ action on Minkowski space. This, in turn, means that at an overlap between two open sets U_i and U_{i+1}, q again transforms as in equation (4.57), which is just the $ISO(2,1)$ version of the general transformation (4.6) for geometric structures. By the definition of page 66, q is thus the developing map, as claimed.

The 'flat coordinate' q appears in the Poincaré gauge theory approach to (2+1)-dimensional gravity, where it is useful in writing down gauge-invariant matter couplings [136, 166]. A construction of this sort has also been used by Newbury and Unruh to investigate exact solutions and the structure of phase space [212].

4.7 The Poisson algebra of the holonomies

We have seen that when (2+1)-dimensional gravity is analyzed in terms of geometric structures or flat connections, the fundamental observables are the holonomies. If we hope to quantize this model, it will therefore be important to understand the Poisson brackets of these quantities, which will eventually become commutators. These are most easily derived from the Poisson brackets in the first-order formalism [205, 209].

For simplicity, I will start with the case of a negative cosmological constant, for which the two $SL(2,\mathbf{R})$ holonomies can be treated independently. Recall from chapter 2 that when $\Lambda = -1/\ell^2$, the two $SL(2,\mathbf{R})$ connections are

$$A^{(\pm)a} = \omega^a \pm \frac{1}{\ell} e^a. \tag{4.58}$$

Using the Poisson brackets (2.100), it is easily checked that

$$\left\{ A_i^{(\pm)a}(x), A_j^{(\pm)b}(x') \right\} = \mp \frac{1}{\ell} \epsilon_{ij} \eta^{ab} \delta^2(x - x'),$$
$$\left\{ A_i^{(+)a}(x), A_j^{(-)b}(x') \right\} = 0. \tag{4.59}$$

Now consider two paths γ_1 and γ_2 intersecting at a point p, as shown in figure 4.5, and let

$$\rho^{\pm}[\gamma_\alpha] = P \exp \left\{ \int_{\gamma_\alpha} A_i^{(\pm)a} T_a dx^i \right\} \tag{4.60}$$

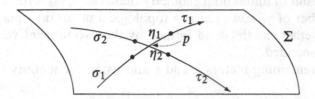

Fig. 4.5. The Poisson bracket of the holonomies receives a contribution only from infinitesimal segments η_1 and η_2 around the intersection point p.

denote the $SL(2, \mathbf{R})$ holonomies (in the gauge-theoretical sense of appendix C) along γ_α. As in chapter 2, the T_a are generators of the Lie algebra of $SL(2, \mathbf{R})$, where in this section we use the two-dimensional representation. The path-ordered exponents (4.60) are commonly known as 'Wilson loops' in gauge theory ('Wilson lines' if γ_α is open), and I will frequently use this terminology. From (4.59), the holonomies have nontrivial brackets only at the point of intersection, and it is convenient to isolate this point. We can do so by writing

$$\gamma_\alpha = \sigma_\alpha \cdot \eta_\alpha \cdot \tau_\alpha, \tag{4.61}$$

where η_1 and η_2 are infinitesimal segments containing p. The product symbol $\cdot$ denotes the product of curves as defined in appendix A; for example, $\sigma \cdot \eta$ is the curve formed by traversing first σ and then η. It follows from the definition of the holonomy that

$$\rho^\pm[\gamma_\alpha] = \rho^\pm[\sigma_\alpha] \cdot \rho^\pm[\eta_\alpha] \cdot \rho^\pm[\tau_\alpha], \tag{4.62}$$

where the product is now the product in $SL(2, \mathbf{R})$.

The only nonvanishing Poisson brackets are those between $\rho[\eta_1]$ and $\rho[\eta_2]$. For such infinitesimal segments, it is sufficient to expand the holonomies around p:

$$\rho^\pm[\eta_\alpha] \sim 1 + \int_{\eta_\alpha} ds A_i^{(\pm)a} T_a \frac{d(x^i \circ \eta_\alpha)}{ds}. \tag{4.63}$$

The brackets (4.59) then give

$$\left\{ \rho^\pm[\eta_1], \rho^\pm[\eta_2] \right\} = \mp \frac{1}{\ell} (\eta^{ab} T_a \otimes T_b) \epsilon(p; \gamma_1, \gamma_2), \tag{4.64}$$

where

$$\epsilon(p; \gamma_1, \gamma_2) \tag{4.65}$$

$$= \int_{s(p)-\epsilon}^{s(p)+\epsilon} ds \int_{s'(p)-\epsilon}^{s'(p)+\epsilon} ds' \epsilon_{ij} \frac{d(x^i \circ \eta_1)}{ds} \frac{d(x^i \circ \eta_2)}{ds'} \delta^2 \left(x \circ \eta_1(s) - x \circ \eta_2(s') \right).$$

It is a standard result of differential geometry that $\epsilon(p; \gamma_1, \gamma_2)$ is the oriented intersection number of γ_1 and γ_2 at p, a topological invariant equal to ± 1 depending on whether the the the dyad formed by the two tangent vectors at p is right- or left-handed.

Restoring the remaining factors σ and τ and using the identity

$$T^{aA}{}_B T_a{}^C{}_D = -\frac{1}{4}\delta^A_B \delta^C_D + \frac{1}{2}\delta^C_B \delta^A_D \qquad (4.66)$$

for the two-dimensional representation of $SL(2, \mathbf{R})$, we obtain

$$\left\{\rho^\pm[\gamma_1]^M{}_N, \rho^\pm[\gamma_2]^P{}_Q\right\} \qquad (4.67)$$

$$= \mp\frac{1}{4\ell}\epsilon(p; \gamma_1, \gamma_2)\left(-\rho^\pm[\gamma_1]^M{}_N \rho^\pm[\gamma_2]^P{}_Q + 2\rho^\pm[\sigma_1 \cdot \tau_2]^M{}_Q \rho^\pm[\sigma_2 \cdot \tau_1]^P{}_N\right)$$

or symbolically,

$$\left\{\rlap{\diagup}\diagdown \,, \right\} = \pm\frac{1}{4\ell}\epsilon(p)\left(\diagdown\!\!\!\diagup -2 \,\right\rangle\!\!\left\langle\,\right). \qquad (4.68)$$

For curves that intersect at more than one point, we must sum a relationship of this form over all intersections.

Let us now consider a set of closed loops γ_α that represent elements of $\pi_1(\Sigma, *)$. The holonomies $\rho^\pm[\gamma_\alpha]$ are not yet gauge-invariant observables: under a gauge transformation

$$A^{(\pm)} \rightarrow g^{(\pm)-1}dg^{(\pm)} + g^{(\pm)-1}Ag^{(\pm)}, \qquad (4.69)$$

we have

$$\rho^\pm[\gamma_\alpha] \rightarrow g^{(\pm)}(*)^{-1}\rho^\pm[\gamma_\alpha]g^{(\pm)}(*), \qquad (4.70)$$

where $*$ is the base point. This noninvariance is a reflection of the equivalence relation (4.21) for holonomies that differ by overall conjugation. The traces

$$R^\pm[\gamma_\alpha] = \frac{1}{2}Tr\rho^\pm[\gamma_\alpha], \qquad (4.71)$$

however, are gauge invariant, and provide an (overcomplete) set of variables on the classical phase space. Their Poisson brackets are easily obtained from (4.67).

In particular, suppose two loops γ_1 and γ_2 intersect only at the base point. The trace of (4.67) then gives

$$\left\{R^\pm[\gamma_1], R^\pm[\gamma_2]\right\} = \mp\frac{1}{4\ell}\epsilon(*; \gamma_1, \gamma_2)\left(R^\pm[\gamma_1 \cdot \gamma_2] - R^\pm[\gamma_1]R^\pm[\gamma_2]\right). \qquad (4.72)$$

For loops with more than one intersection, the algebra is somewhat more complicated; if the intersections are at points p_i, equation (4.72) must be replaced by

$$\left\{R^{\pm}[\gamma_1], R^{\pm}[\gamma_2]\right\} = \mp\frac{1}{4\ell}\sum_i \epsilon(p_i; \gamma_1, \gamma_2)\left(R^{\pm}[\gamma_1 \cdot_i \gamma_2] - R^{\pm}[\gamma_1]R^{\pm}[\gamma_2]\right).$$

(4.73)

In this expression, the symbol $\cdot_i$ denotes the product of curves, as described in appendix A, with p_i treated as the base point; that is, $\gamma_1 \cdot_i \gamma_2$ is the loop obtained by starting at p_i and first traversing γ_1, then γ_2. Because of this composition, it is difficult to find small subalgebras of the algebra of holonomies that close under Poisson brackets. However, Nelson and Regge have succeeded in constructing a small but complete (actually overcomplete) set of holonomies on a surface of arbitrary genus that form a closed algebra [206, 207, 208].

For the torus, in particular, the geometric structure is completely determined by a closed subalgebra of three holonomies. Let γ_1 and γ_2 be two independent circumferences, and denote the traces $R^{\pm}[\gamma_1]$, $R^{\pm}[\gamma_2]$, and $R^{\pm}[\gamma_1 \cdot \gamma_2]$ as $R_1^{\pm}$, $R_2^{\pm}$, and $R_{12}^{\pm}$. We then have

$$\left\{R_1^{\pm}, R_2^{\pm}\right\} = \mp\frac{4\pi G}{\ell}(R_{12}^{\pm} - R_1^{\pm}R_2^{\pm}) \quad \textit{and cyclical permutations,}$$

(4.74)

where I have restored the gravitational constant. The six traces $R_\alpha^{\pm}$ are, in fact, overcomplete – the space of geometric structures on the torus is only four-dimensional. To understand this overcompleteness, consider the cubic polynomials

$$F^{\pm} = 1 - (R_1^{\pm})^2 - (R_2^{\pm})^2 - (R_{12}^{\pm})^2 + 2R_1^{\pm}R_2^{\pm}R_{12}^{\pm}$$

$$= \frac{1}{2}Tr\left(I - \rho^{\pm}[\gamma_1]\rho^{\pm}[\gamma_2]\rho^{\pm}[\gamma_1^{-1}]\rho^{\pm}[\gamma_2^{-1}]\right),$$

(4.75)

where the last equality follows from the identity

$$A + A^{-1} = I\, Tr\, A$$

(4.76)

for matrices $A \in SL(2, \mathbf{R})$. The $F^{\pm}$ vanish classically, since the fundamental group of the torus satisfies the relation

$$\gamma_1 \cdot \gamma_2 \cdot \gamma_1^{-1} \cdot \gamma_2^{-1} = I,$$

(4.77)

and the conditions $F^{\pm} = 0$ thus provide two relations among the six $R_\alpha^{\pm}$.

For the torus, we have already obtained an explicit representation (4.33) for the holonomies in terms of a set of coordinates $r_a^{\pm}$ on the space of geometric structures. It is now fairly easy to show that the symplectic structure (4.39) for these coordinates is equivalent to the symplectic structure of equation (4.74), and that the polynomials $F^\pm$ vanish when expressed in these coordinates, confirming the consistency of our descriptions.

Although the brackets (4.73) were derived in the context of general relativity, it is interesting to note that they occur as well in the general theory of Riemann surfaces. The moduli space

$$\widetilde{\mathcal{N}} = Hom_0(\pi_1(\Sigma), SO(2,1))/\sim \tag{4.78}$$

of section 3 admits a natural symplectic structure, and in investigating that structure Goldman has found a set of Poisson brackets equivalent to those of (2+1)-dimensional gravity [129]. A similar set of relations occurs when one considers homomorphisms from $\pi_1(\Sigma)$ into an arbitrary group G, although for $SL(2,\mathbf{R})$ an additional set of relations holds: it follows from (4.76) that $Tr\,AB + Tr\,AB^{-1} = Tr\,A\,Tr\,B$, and hence

$$R^\pm[\gamma_1]R^\pm[\gamma_2] = \frac{1}{2}\left(R^\pm[\gamma_1 \cdot \gamma_2] + R^\pm[\gamma_1 \cdot \gamma_2^{-1}]\right). \tag{4.79}$$

Relations of this type are known as Mandelstam constraints.

In the case of a vanishing cosmological constant, a similar algebra can be found, using an appropriate matrix representation of $ISO(2,1)$ [188]. It is often more useful to choose a slightly different set of variables, however, to make use of the cotangent bundle structure of the space of geometric structures. (Recall that this structure occurred only for $\Lambda = 0$.) Let

$$\rho_0[\gamma, x] = P\exp\left\{\int_\gamma \omega^a \mathscr{J}_a\right\} \tag{4.80}$$

be the holonomy of the $SO(2,1)$ connection one-form ω^a around a loop γ, with base point x. As before, the $\mathscr{J}_a$ are a set of generators of the Lie algebra of $SO(2,1)$; it is traditional to use the two-dimensional representation of $SU(2)$, the covering space of $SO(2,1)$. The Ashtekar–Rovelli–Smolin loop variables are then [11, 13]

$$T^0[\gamma] = \frac{1}{2}Tr\,\rho_0[\gamma, x] \tag{4.81}$$

and

$$T^1[\gamma] = \int_\gamma Tr\left\{\rho_0[\gamma, x(s)]\,e^a(\gamma(s))\mathscr{J}_a\right\}. \tag{4.82}$$

$T^0[\gamma]$ may be recognized as the $ISO(2,1)$ holonomy in the representation $\mathscr{P}^a = 0$. $T^1[\gamma]$ is essentially a cotangent vector to $T^0[\gamma]$. Indeed, given a family $\omega^a(t)$ of flat connections, the derivative of (4.81) yields

$$2\frac{d}{dt}T^0[\gamma](t) = \int_\gamma Tr\left(\rho_0(t)[\gamma, x(s)]\frac{d}{dt}\omega^a(t)(\gamma(s))\mathcal{J}_a\right),$$
(4.83)

and we saw in chapter 2 that triads e^a may be interpreted as cotangent vectors $d\omega^a/dt$ to the space of flat connections.

The variables T^0 and T^1 obey a closed Poisson algebra. The computation of the relevant brackets is nearly identical to the computation for $SL(2,\mathbf{R})$ described above; for a set of intersection points p_i, we obtain

$$\left\{T^0[\gamma_1], T^0[\gamma_2]\right\} = 0$$

$$\left\{T^1[\gamma_1], T^0[\gamma_2]\right\} = -\frac{1}{8}\sum_i \epsilon(p_i; \gamma_1, \gamma_2)\left(T^0[\gamma_1 \cdot_i \gamma_2] - T^0[\gamma_1 \cdot_i \gamma_2^{-1}]\right)$$

$$\left\{T^1[\gamma_1], T^1[\gamma_2]\right\} = -\frac{1}{8}\sum_i \epsilon(p_i; \gamma_1, \gamma_2)\left(T^1[\gamma_1 \cdot_i \gamma_2] - T^1[\gamma_1 \cdot_i \gamma_2^{-1}]\right),$$
(4.84)

where the product $\cdot_i$ was defined on page 83. $T^0[\gamma]$ and $T^1[\gamma]$ also obey a set of Mandelstam constraints analogous to (4.79):

$$T^0[\gamma_1]T^0[\gamma_2] = \frac{1}{2}\left(T^0[\gamma_1 \cdot \gamma_2] + T^0[\gamma_1 \cdot \gamma_2^{-1}]\right)$$

$$T^0[\gamma_1]T^1[\gamma_2] + T^0[\gamma_2]T^1[\gamma_1] = \frac{1}{2}\left(T^1[\gamma_1 \cdot \gamma_2] + T^1[\gamma_1 \cdot \gamma_2^{-1}]\right),$$
(4.85)

along with the identities

$$T^0[0] = 1, \qquad T^0[\gamma] = T^0[\gamma^{-1}], \qquad T^0[\gamma_1 \cdot \gamma_2] = T^0[\gamma_2 \cdot \gamma_1]$$
$$T^1[0] = 0, \qquad T^1[\gamma] = T^1[\gamma^{-1}], \qquad T^1[\gamma_1 \cdot \gamma_2] = T^1[\gamma_2 \cdot \gamma_1]. \quad (4.86)$$

Once again, the composition of loops in (4.84) and (4.85) makes it difficult to extract a finite-dimensional set of observables that parametrize the phase space $T^*\mathcal{N}$. As usual, however, if Σ has the topology of a torus, the mathematics simplifies drastically. As discussed in appendix A, the two generators γ_1 and γ_2 of $\pi_1(T^2)$ commute, so any curve on the torus is homotopic to a curve $\gamma_1{}^m \cdot \gamma_2{}^n$. Any homotopy class may thus be labeled by two integers m and n, the winding numbers in the x and y directions. We may therefore think of T^0 and T^1 as being functions of these two integers.

Moreover, using the explicit expressions (3.76) and (3.78) for the triad and spin connection on $[0,1] \times T^2$, we can compute these functions explicitly. We find

$$T^0[m, n] = \cosh \frac{m\lambda + n\mu}{2}, \qquad T^1[m, n] = -(ma + nb)\sinh \frac{m\lambda + n\mu}{2}.$$
(4.87)

The functions T^0 and T^1 clearly determine λ, μ, a, and b, and can thus be used to label points in the space $\mathcal{M}$ of classical solutions. Moreover, it is now an easy exercise to check explicitly that the Poisson brackets (4.84) of the holonomy variables imply the symplectic structure (3.82) derived in chapter 3, once again confirming the consistency of our various descriptions.

For spatial topologies of genus $g > 1$, no simple expression analogous to (4.87) is known. However, Loll has recently found a complete set of independent loop variables sufficient to parametrize the full phase space $T^*\mathcal{N}$ [178]. Loll's $6g - 6$ T^0 variables are written in terms of Fenchel–Nielsen coordinates, a set of coordinates on higher-genus moduli space analogous to the modulus τ of the torus; her conjugate T^1 variables are expressed as differential operators that satisfy the appropriate commutation relations.

5

Canonical quantization in reduced phase space

Having examined the classical dynamics of (2+1)-dimensional gravity, we are now ready to turn to the problem of quantization. As we shall see in the next few chapters, there are a number of inequivalent approaches to quantum gravity in 2+1 dimensions. In particular, each of the the classical formalisms of the preceding chapters – the ADM representation, the Chern–Simons formulation, the method of geometric structures – suggests a corresponding quantum theory.

The world is not (2+1)-dimensional, of course, and the quantum theories developed here cannot be taken too literally. Our goal is rather to learn what we can about general features of quantum gravity, in the hope that these lessons may carry over to 3+1 dimensions. Fortunately, many of the basic conceptual issues of quantum gravity do not depend on the number of dimensions, so we might reasonably hope that even a relatively simple model could provide useful insights.

After a brief introduction to some of the conceptual issues we will face, I will devote this chapter to a quantum theory based on the ADM representation of chapter 2. As we saw in that chapter, the ADM decomposition and the York time-slicing make it possible to reduce (2+1)-dimensional gravity to a system of finitely many degrees of freedom. Quantum gravity thus becomes quantum mechanics, a subject we believe we understand fairly well. This approach has important limitations, which are discussed at the end of this chapter, but it is a good starting place.

5.1 Conceptual issues in quantum gravity

The assumption of a fixed spacetime background pervades ordinary quantum theory. It appears in the definition of equal time commutators, in the normalization of wave functions on spacelike surfaces, in the imposition of causality requirements, even in the choice of fundamental observables.

87

In general relativity, on the other hand, the universe is dynamical, and quantum gravity requires quantization of the structure of spacetime itself. It should come as no surprise that attempts to formulate such a theory quickly bring deep conceptual issues to the fore.

One serious difficulty arises as soon as an attempt is made to determine the quantum mechanical observables. A coordinate system in general relativity has no objective physical meaning – the theory is invariant under diffeomorphisms – and physical observables must therefore be independent of the choice of coordinates. As we saw in chapter 2, the diffeomorphisms are generated by the Hamiltonian and momentum constraints in classical general relativity. In the quantum theory, Poisson brackets become commutators, and diffeomorphism invariance becomes the requirement that observables commute with the constraints and that physical states be annihilated by them. So far, however, it has proven difficult to find *any* observables that meet this requirement, much less a complete set. In particular, observables cannot be local functions of the coordinates, since diffeomorphisms shuffle points around but must leave observables unchanged.

A particularly strong form of this dilemma goes under the name of the 'problem of time'. (For good reviews, see references [160] and [171].) In conventional approaches to canonical quantization, observables in quantum gravity can no more depend on time than on the spatial coordinates: the Hamiltonian constraint generates translations in coordinate time, and must commute with observables and annihilate physical states.* Observables must therefore be time-independent – that is, they must be constants of motion! On the other hand, general relativity is clearly a dynamical theory, and quantum gravity must somehow describe this dynamics. One may try to circumvent this problem by choosing a 'physical time', for instance by slicing spacetime into hypersurfaces of constant mean extrinsic curvature. But such a selection seems arbitrary, and the resulting quantum theory is likely to depend on the choice of slicing. One may try to build physical 'clocks' out of matter, but two different clocks will not always agree: any clock constructed from a quantum mechanical field with a nonnegative Hamiltonian has a finite probability of occasionally running backwards [261]. Alternatively, one may try to single out the Hamiltonian constraint for special treatment, using it as an equation of motion rather than a restriction on physical states, but this seems to violate at least the spirit of general covariance.

Even if one could find a complete set of physical observables, it would remain necessary to interpret their values in terms of information about

---

* Strictly speaking, we saw in chapter 2 that the Hamiltonian constraint (2.13) generates diffeomorphisms only on shell, creating a further difficulty in interpretation.

geometry. In a sense, this is equivalent to completely integrating the classical equations of motion: one would have to reconstruct the geometry of spacetime from a set of diffeomorphism-invariant constants of motion. Needless to say, this is not an easy task.

Past attempts to solve these and similar problems have given birth to a plethora of techniques. One may start with the path integral or with canonical quantization. In the path integral approach, one may fix the topology of spacetime or sum over some or all topologies. In canonical quantization, one may solve the constraints classically or impose them as conditions on physical states. In either approach, one may choose among a wide assortment of fundamental variables, and given a choice of phase space variables, one may further select among 'polarizations' into positions and momenta. In 3+1 dimensions, it is not clear that any of these approaches works, and it is certainly not obvious which of them, if any, are equivalent.

One long-term aim of (2+1)-dimensional gravity is to sort out some of these conceptual issues. Beyond this, we can also explore some of the qualitative features that might be expected in a more realistic theory of quantum gravity. Can the topology of space change in time? What is the effect of summing over spacetime topologies? Can gravity cut off the ultraviolet divergence of quantum field theory? Why is the cosmological constant so nearly zero? Do quantum effects prevent the formation of closed timelike curves and 'time machines'?

5.2 Quantization of the reduced phase space

The starting point of any quantum theory is a classical formulation to be 'quantized'. This is perhaps unfortunate – presumably the quantum theory should be fundamental, and the classical theory derived – but we do not know how to proceed otherwise.

A simple starting point for quantum gravity in 2+1 dimensions is the ADM formalism of chapter 2 in the York time-slicing. We begin by assuming that spacetime has the (fixed) topology $[0,1] \times \Sigma$, where Σ is a closed surface of genus $g > 0$. As we saw in chapter 2, the dynamics of the physical degrees of freedom is described by a reduced phase space action

$$I = \int dT \left\{ p^\alpha \frac{dm_\alpha}{dT} - H_{red}(m, p, T) \right\}, \tag{5.1}$$

where

$$H_{red} = \int_{\Sigma_T} d^2x \sqrt{\tilde{g}} \, e^{2\lambda(m,p,T)} \tag{5.2}$$

and λ is a complicated function of m, p, and T obtained by solving the elliptic differential equation (2.35). The positions m_α parametrize the moduli space $\mathcal{N}$ of Σ, and the momenta p^α are cotangent vectors, so the ADM phase space has the structure of a cotangent bundle $T^*\mathcal{N}$.

In principle, the quantization of the action (5.1) is straightforward. We simply promote the m_α and p^α to operators on the Hilbert space of square-integrable functions on $\mathcal{N}$, with equal time commutation relations

$$[\hat{m}_\alpha, \hat{p}^\beta] = i\delta_\alpha^\beta, \tag{5.3}$$

i.e.,

$$\hat{p}^\alpha = -i\frac{\partial}{\partial m_\alpha}. \tag{5.4}$$

The resulting quantum mechanical system then has simple states, but the dynamics is extremely complicated. As discussed in chapter 3, for spatial topologies of genus $g > 1$ it is not clear that equation (2.35) can be solved for λ in closed form, although there is some hope of finding the exact solution for genus 2. But even if a solution can be found, the resulting Hamiltonian will be an explicitly time-dependent, nonpolynomial function of both coordinates and momenta, leading to a highly ambiguous operator ordering.

As in the classical analysis, however, these difficulties largely disappear if Σ has the topology of a torus. We can then use the Hamiltonian of equation (3.66) to write down a Schrödinger equation[†]

$$i\frac{\partial}{\partial T}\psi(\tau, \bar{\tau}, T) = \hat{H}_{red}\psi(\tau, \bar{\tau}, T) \tag{5.5}$$

with

$$\hat{H}_{red} = (T^2 - 4\Lambda)^{-1/2}\hat{\tau}_2\left((\hat{p}^1)^2 + (\hat{p}^2)^2\right)^{1/2} = (T^2 - 4\Lambda)^{-1/2}\Delta_0^{1/2}. \tag{5.6}$$

Here

$$\Delta_0 = -\tau_2^2\left(\frac{\partial^2}{\partial\tau_1^2} + \frac{\partial^2}{\partial\tau_2^2}\right) \tag{5.7}$$

is the standard scalar Laplacian for the metric (3.67) on moduli space,

$$ds_\mathcal{N}^2 = \frac{1}{\tau_2^2}d\tau d\bar{\tau}. \tag{5.8}$$

[†] A 'Klein–Gordon' version of this equation, with T replaced by its conjugate variable V, was found by Martinec as early as 1984 [189].

This metric has constant negative curvature, and is defined on the upper half-plane $\tau_2 > 0$; we saw in chapter 3 that the classical solutions were geodesics with respect to this geometry. Recall also that the large diffeomorphisms generate an additional symmetry, that of the mapping class (or modular) group, generated by the transformations

$$S : \tau \to -\frac{1}{\tau}$$

$$T : \tau \to \tau + 1, \tag{5.9}$$

which leave the Laplacian (5.7) invariant. Classical observables are invariant under these transformations, which are, after all, merely diffeomorphisms; it is reasonable to assume that wave functions should also be invariant, or at least covariant.

The Laplacian acting on such invariant functions is known to mathematicians as the Maass Laplacian of weight zero, and its eigenfunctions are weight zero Maass forms [111, 222, 226, 255]. It may be shown that Δ_0 is self-adjoint and has a nonnegative spectrum, so the square root in equation (5.6) can be defined by the spectral decomposition. That is, any modular invariant function ψ can be expanded in terms of eigenfunctions of Δ_0,

$$\psi(\tau, \bar{\tau}) = \sum_v c_v \psi_v(\tau, \bar{\tau}), \tag{5.10}$$

where the ψ_v are weight zero Maass forms,

$$\Delta_0 \psi_v(\tau, \bar{\tau}) = \lambda_v \psi_v(\tau, \bar{\tau}), \quad \psi_v(\tau+1, \bar{\tau}+1) = \psi_v(-\frac{1}{\tau}, -\frac{1}{\bar{\tau}}) = \psi_v(\tau, \bar{\tau}). \tag{5.11}$$

We can define the action of $\Delta_0^{1/2}$ on ψ as

$$\Delta_0^{1/2} \psi(\tau, \bar{\tau}) = \sum_v c_v \lambda_v^{1/2} \psi_v(\tau, \bar{\tau}). \tag{5.12}$$

A general time-dependent wave function – a solution of equation (5.5) – then takes the form

$$\psi(\tau, \bar{\tau}, T) = \sum_v c_v e^{-i\lambda_v^{1/2} t} \psi_v(\tau, \bar{\tau}), \tag{5.13}$$

where t is the time coordinate of equation (3.68),

$$t = \int \frac{dT}{\sqrt{T^2 - 4\Lambda}}. \tag{5.14}$$

The eigenfunctions of the Laplacian (5.7) can be found explicitly by separation of variables. One set of eigenfunctions is

$$u_v^{(n)}(\tau_1, \tau_2) = \tau_2^{1/2} K_{iv}(2\pi|n|\tau_2)e^{2\pi i n \tau_1}, \tag{5.15}$$

where K_{iv} is a modified Bessel function. The eigenfunction $u_v^{(n)}$ has eigenvalue $v^2 + \frac{1}{4}$, independent of n, and if n is an integer, $u_v^{(n)}$ is clearly invariant under the transformation $T : \tau_1 \to \tau_1 + 1$. An additional set of eigenfunctions with the same eigenvalues is

$$h_v^{(\pm)}(\tau_1, \tau_2) = \tau_2^{1/2} e^{\pm iv \ln \tau_2}, \tag{5.16}$$

and the functions $h_v^{(\pm)}$ are trivially T-invariant. The $u_v^{(n)}$ and $h_v^{(\pm)}$ are not, however, invariant under the transformation $S : \tau \to -1/\tau$. A weight zero Maass form must therefore be a superposition

$$\psi_v(\tau, \bar\tau) = \sum_n \rho_v(n) u_v^{(n)} + \rho_v^+ h_v^+ + \rho_v^- h_v^-. \tag{5.17}$$

Modular invariant superpositions of the form (5.17) exist for all values of v. The continuous part of the spectrum of Δ_0 comes from a collection of eigenfunctions $E_v(\tau_1, \tau_2)$ known as Eisenstein series, whose coefficients ρ_v are known explicitly. Additional eigenfunctions $v_n(\tau_1, \tau_2)$, known as cusp forms, occur for particular values of v, adding a discrete component to the spectrum. The corresponding coefficients ρ_n and eigenvalues λ_n are known only numerically. The cusp forms are square integrable with respect to the measure determined by the metric (5.8), while the $E_v(\tau_1, \tau_2)$ are not. In general, however, both are needed in the expansion of an arbitrary modular invariant function [255]: the Roelcke–Selberg spectral decomposition theorem states that any modular invariant square integrable function $\psi(\tau, \bar\tau)$ has an expansion

$$\psi(\tau, \bar\tau) = \sum \langle v_n | \psi \rangle v_n(\tau, \bar\tau) + \frac{1}{4\pi} \int dv \langle E_v | \psi \rangle E_v(\tau, \bar\tau) \tag{5.18}$$

where

$$\langle f | \psi \rangle = \int_{\mathcal{N}} \frac{d^2\tau}{\tau_2{}^2} f^*(\tau, \bar\tau)\psi(\tau, \bar\tau). \tag{5.19}$$

Equation (5.18) may be understood as the proper form of the expansion (5.10), with both the discrete and the continuous spectra taken into account.

The requirement of modular invariance introduces a rather peculiar feature to the quantum theory. Since the modulus τ is not itself invariant, $\hat\tau$ is no longer, strictly speaking, an admissible observable: that is, if $\psi(\tau, \bar\tau)$

is an invariant wave function, $\tau\psi(\tau,\bar{\tau})$ is not. A similar phenomenon occurs in the quantization of a particle moving on a circle: the angle θ is not itself periodic, and cannot be made into a self-adjoint operator. This is not a serious problem, of course, since the trigonometric functions $\cos\theta$ and $\sin\theta$ are perfectly good operators, which together give us exactly the information we need from θ. Similarly, for the model of quantum gravity developed here we can obtain complete information about $\hat{\tau}$ from an appropriate set of modular invariant operators. The analogs of the trigonometric functions are now the 'modular functions', which have been studied extensively by mathematicians. A standard example is the modular function $J(\tau)$ of Dedekind and Klein, defined by [225]

$$J(\tau) = \frac{(60G_4(\tau))^3}{(60G_4(\tau))^3 - 27(140G_6(\tau))^2}, \tag{5.20}$$

where the $G_{2k}(\tau)$ are Eisenstein series,

$$G_{2k}(\tau) = \sum_{m,n\in\mathbf{Z}}' \frac{1}{(m+n\tau)^{2k}}. \tag{5.21}$$

(The prime means that the value $m = n = 0$ is excluded from the sum.) It may be shown that any meromorphic modular function is a rational function of $J(\tau)$. Such functions are certainly less familiar than trigonometric functions, but in principle they are no more extraordinary.

5.3 Automorphic forms and Maass operators

The quantum theory described above was based on wave functions that were invariant under the action (5.9) of the mapping class group. This may be too strong a requirement, however: it should be sufficient to demand that wave functions transform under a unitary representation, so that the inner products that give transition amplitudes are invariant. To understand the range of possible quantum theories, we must take a short digression into the theory of representations of the torus mapping class group. This group is also known as the modular group, and it is represented on spaces of functions known as automorphic forms (strictly speaking, 'automorphic forms for the modular group', or in the case of holomorphic representations, 'modular forms'). The study of automorphic forms is a major topic in mathematics, and we shall only touch on the highlights here; for more detail, see, for example, reference [255].

Recall first that the modulus τ of the torus lies in the upper half-plane $\tau_2 > 0$. This is the Teichmüller space $\widetilde{\mathcal{N}}(T^2)$ of the torus; more generally, the space of constant curvature metrics on a surface Σ modulo small diffeomorphisms is the Teichmüller space $\widetilde{\mathcal{N}}(\Sigma)$, as discussed in appendix A.

The mapping class group – the group of large diffeomorphisms – acts on $\widetilde{\mathscr{N}}(\Sigma)$, and the quotient by this action is the moduli space $\mathscr{N}(\Sigma)$.

The mapping class group of the torus is the group generated by the transformations S and T of equation (5.9). An arbitrary element of this group may be represented by a unimodular matrix of integers

$$\gamma = \begin{pmatrix} a & b \\ c & d \end{pmatrix}, \qquad a,b,c,d \in \mathbf{Z}, \quad \det \begin{vmatrix} a & b \\ c & d \end{vmatrix} = 1 \qquad (5.22)$$

with the corresponding transformation

$$\tau \mapsto \gamma\tau = \frac{a\tau + b}{c\tau + d}, \qquad (5.23)$$

so

$$S = \begin{pmatrix} 0 & 1 \\ -1 & 0 \end{pmatrix}, \qquad T = \begin{pmatrix} 1 & 1 \\ 0 & 1 \end{pmatrix}. \qquad (5.24)$$

The composition of two such transformations corresponds to matrix multiplication, and it may be checked that the upper half-space metric (5.8) is invariant. We shall also need the transformation of the differential $d\tau$,

$$d\tau \mapsto (c\tau + d)^{-2}d\tau, \qquad (5.25)$$

and of τ_2,

$$\tau_2 \mapsto |c\tau + d|^{-2}\tau_2. \qquad (5.26)$$

The matrices (5.22) consitute the group $SL(2,\mathbf{Z})$. Note, though, that the transformation (5.23) remains the same if we simultaneously change the signs of a, b, c, and d. The true torus mapping class group is thus $SL(2,\mathbf{Z})/\{I,-I\} \approx PSL(2,\mathbf{Z})$.

Now let us return to the question of how wave functions ought to transform under the modular group. We know of one obvious way to generalize ordinary functions on Teichmüller space: we can consider wave functions that are spinors or tensors. For a Teichmüller space of greater than two dimensions, such objects would have many components, and would give us multidimensional representations of the mapping class group. For the torus, however, $\mathscr{N}$ is a two-dimensional space with a natural complex structure, and it is possible to define one-component holomorphic and antiholomorphic spinors and tensors.[‡]

[‡] Readers unfamiliar with the representation of two-dimensional spinors as forms of half-integral weight might wish to consult reference [2], which also contains an interesting introduction to Teichmüller spaces.

An *automorphic form* (for the modular group) of weight (p,q), with p and q integers, is defined to be a function $f(\tau)$ that satisfies

$$f(\gamma\tau) = (c\tau + d)^p(c\bar{\tau} + d)^q f(\tau) \tag{5.27}$$

for any transformation γ in the torus mapping class group. Such a function has a natural interpretation as a differential form on $\widetilde{\mathcal{N}}$ that projects down to $\mathcal{N}$, since the combination

$$f(\tau)d\tau^{p/2}d\bar{\tau}^{q/2}$$

is invariant. If p and q are even, an automorphic form of weight (p,q) is an ordinary tensor of rank $(p/2, q/2)$, while if p or q is odd, f is a spinor. By (5.26), we can convert an automorphic form of weight (p,q) to one of weight $((p-q)/2, -(p-q)/2)$ by multiplying by a suitable power of τ_2. It is customary to do so, and to consider only forms of weight $(k,-k)$, which are commonly referred to simply as 'forms of weight k'. Note that k can now be an integer or a half-integer. Indeed, an ordinary spinor (weight $(1,0)$) corresponds to a value $k = 1/2$, and quite generally a form of half-integral weight will have spinorial transformation properties.

For p or q odd, the transformation (5.27) is somewhat ambiguous, since it is not invariant under $(c,d) \mapsto (-c,-d)$. A similar ambiguity can arise from the need to multiply by half-integral powers of τ_2. The complete resolution of these problems is somewhat complicated, and I shall not discuss it here (see, for instance, reference [255]); let me merely note that for an automorphic form of weight k, we must require that

$$f(\gamma\tau) = e^{-2\pi ik}f(\tau) \qquad \text{for } \gamma = -I. \tag{5.28}$$

Note also that under the action of the generator S, a form of weight k transforms as

$$f(S\tau) = \left(\frac{\tau}{\bar{\tau}}\right)^k f(\tau). \tag{5.29}$$

We shall use this result later when we compare reduced phase space ADM quantization to Chern–Simons quantization.

Now let $\mathcal{F}^n$ be the space of finite-norm forms of weight $(n,-n)$. To differentiate such forms, we introduce the *Maass operators*

$$K_n = (\tau - \bar{\tau})\frac{\partial}{\partial\tau} + n \quad : \quad \mathcal{F}^n \to \mathcal{F}^{n+1} \tag{5.30}$$

and

$$L_n = (\bar{\tau} - \tau)\frac{\partial}{\partial\bar{\tau}} - n \quad : \quad \mathcal{F}^n \to \mathcal{F}^{n-1}. \tag{5.31}$$

These are essentially covariant derivatives with respect to the metric (5.8). The corresponding Laplacians are

$$\Delta_n = -L_{n+1}K_n \tag{5.32}$$

$$= -\tau_2{}^2 \left(\frac{\partial^2}{\partial \tau_1{}^2} + \frac{\partial^2}{\partial \tau_2{}^2} \right) + 2in\tau_2 \frac{\partial}{\partial \tau_1} + n(n+1) \quad : \quad \mathscr{F}^n \to \mathscr{F}^n$$

and

$$-K_{n-1}L_n = \Delta_n - 2n. \tag{5.33}$$

In particular, the Maass Laplacian for the space $\mathscr{F}^0$ of forms of weight zero is just the ordinary Laplacian (5.7).

The space $\mathscr{F}^n$ has a natural inner product (compare (5.19)):

$$\langle f_1 | f_2 \rangle = \int_{\mathscr{N}} \frac{d^2\tau}{\tau_2{}^2} f_1^*(\tau, \bar{\tau}) f_2(\tau, \bar{\tau}), \tag{5.34}$$

which can be obtained from the constant negative curvature metric (5.8). It is easily checked that this product is invariant under the modular group. The Maass operators K and L are adjoints with respect to this inner product,

$$L_n{}^\dagger = -K_{n-1}, \tag{5.35}$$

and the Laplacian (5.32) is self-adjoint.

5.4 A general ADM quantization

The general form of the quantization introduced in section 2 is now clear. As our Hilbert space, we take the space $\mathscr{F}^n$ of finite-norm automorphic forms of weight n, with the inner product (5.34). As in section 2, the modulus τ is not an operator on $\mathscr{F}^n$, but modular functions can again be made into operators that act by multiplication. Similarly, the partial derivative with respect to τ is not an operator on $\mathscr{F}^n$, but the operators K_n and L_n act as covariant derivatives, from which momentum operators can be constructed. For example, if $f(\tau)$ is a form of weight 1, then the operator

$$\hat{p}_f = f(\tau)L_n \quad : \quad \mathscr{F}^n \to \mathscr{F}^n \tag{5.36}$$

is a 'derivative in the f direction'.

For our 'Schrödinger equation', we can now take the obvious generalization of (5.5):

$$i\frac{\partial}{\partial T}\psi(\tau, \bar{\tau}, T) = \hat{H}_{red}^{(n)}\psi(\tau, \bar{\tau}, T) \tag{5.37}$$

with

$$\hat{H}_{red}^{(n)} = (T^2 - 4\Lambda)^{-1/2}\Delta_n^{1/2}, \tag{5.38}$$

where Δ_n is the Maass Laplacian (5.32). This Hamiltonian differs from our original expression (5.6) by terms of order $\hbar$, and can be viewed as a different operator ordering of the classical Hamiltonian. Quantum theories with different values of n are clearly distinct: the eigenvalues of the Hamiltonian $\hat{H}_{red}^{(n)}$, for example, depend on n. There seems to be no *a priori* reason for choosing one value over another. In the ADM formalism, for example, the choice $n = 0$ may seem most natural, but we shall see in chapter 6 that the first-order formalism seems to lead most naturally to a choice $n = 1/2$.

With hindsight, the appearance of many inequivalent quantizations should come as no surprise. A classical theory determines the corresponding quantum theory only up to choices of operator ordering, and different orderings can lead to physically inequivalent theories. Ultimately, such ambiguities must be resolved by observation. For example, much of 'old quantum mechanics', which we now understand to be incorrect, can be rephrased in modern language as a 'wrong' choice of variables – namely, action and angle variables – upon which to impose canonical commutation relations. When these variables are reexpressed in terms of the 'right' ps and qs, differences in operator ordering lead to physical predictions that disagree with experiment. Both choices have the same classical limit; it is only comparison with the real world that shows us which quantization is 'right'.

From this point of view, the role of the mapping class group can be compared to that of the Poincaré group in ordinary quantum field theory. The existence of such a group does not determine the quantum theory, even at the kinematical level, but it does greatly restrict the possibilities. The different weights are roughly analogous to the different spins that can occur in particle physics.

5.5 Pros and cons

Do we now have a theory of (2+1)-dimensional quantum gravity? The procedure of the preceding sections is fairly straightforward, at least in principle, and it gives us a set of models that are, at least, not obviously wrong. Nevertheless, several problems must be considered before we accept these models as the last word.

The first problem is, in a sense, technical. The Hamiltonian obtained above came from the solution of the complicated differential equation (2.35). For spatial topologies of genus greater than one, we do not know how· to solve this equation. In itself, this is not so terrible – we can look for approximate solutions, for example, and try to develop a perturbation theory for the Hamiltonian. The problem is that the result will almost certainly be a complicated, time-dependent function of both positions and momenta, with terrible operator ordering ambiguities. In the genus one case, the mapping class group dramatically reduced such ambiguities, but there is no guarantee that anything so simple will happen for more complicated topologies. It seems particularly unlikely that any such simplification will occur order by order in a perturbation expansion.

A second, related problem comes from the presence of a square root in the Hamiltonians (5.6) and (5.38). In the eigenvalue expansion (5.12), I implicitly assumed that the relevant operator was the positive square root. Classically, it is evident from the equations of motion (3.69) that a change in the sign of this square root corresponds to a reversal of the momenta p_1 and p_2, that is, a switch from an expanding to a collapsing universe. Quantum mechanically, however, there seems to be no reason to choose the same sign for each mode in (5.12): we could have a wave function describing an arbitrary mixture of expanding and collapsing modes. We must make a choice, but there seems to be nothing in the formalism to tell us what choice to make. One possible resolution is to note from equation (2.37) that the classical reduced Hamiltonian is an area, and should therefore presumably be positive; it is not clear whether this argument should carry over to the quantum theory.

A third problem, perhaps the most serious, comes from the treatment of time-slicing. The choice of York time, $TrK = -T$, was made *classically*, and the quantum theories of sections 2 and 4 are based on this choice. Such a procedure violates at least the spirit of general covariance, which states that no choice of a time coordinate is preferable to any other. It should be stressed that this classical choice does not mean that the probability amplitudes computed in these models are unphysical. The statement, 'The spacelike hypersurface of mean curvature $TrK = -3$ has modulus $1 + i$' is an invariant claim about a classical geometry, and its quantum mechanical counterpart should be physically meaningful. The problem is that there are other, equally meaningful statements – for example, 'The hypersurface of constant *intrinsic* curvature -3 has modulus $1 + i$' – whose expressions in this formalism are, at the least, obscure.

This problem could perhaps be resolved by looking at quantum theories based on different choices of classical time-slicing. We saw in chapter 2 that the choice $TrK = -T$ is particularly simple, but there is nothing in principle to prevent us from performing a similar reduction to the physical degrees of freedom with a different choice of time. Classically, all such reduced phase space theories are equivalent. Quantum mechanically, however, it is not at all obvious that different classical choices lead to equivalent predictions.

In particular, consider a section of spacetime, say $[0,1] \times \Sigma$, with fixed values of TrK on the initial and final surfaces $\{0\} \times \Sigma$ and $\{1\} \times \Sigma$ and a fixed wave function $\psi(\tau, \bar{\tau}, 0)$ on the initial slice. The models developed in this chapter will predict a definite final wave function $\psi(\tau, \bar{\tau}, 1)$. But there are other choices of time-slicing that agree with the TrK slicing at the initial and final surfaces but disagree in the interior, and it is not obvious that the quantum theories based on these slicings will lead to the same $\psi(\tau, \bar{\tau}, 1)$. Cosgrove has argued that such differences can very probably be absorbed in the choice of operator ordering of the Hamiltonian [81]. That is, there are (plausibly) choices of operator orderings that will make any two intermediate slicings – or, in fact, infinitely many intermediate slicings – agree. Once again, however, the orderings are not unique, and it is not clear that any should be preferred; the 'natural' ordering in one time-slicing may seem highly 'unnatural' in another. We shall return to this question in the next chapter, after we have developed a 'covariant canonical quantum theory' that does not depend on a classical choice of time.

6

The connection representation

The quantum theory of the preceding chapter grew out of the ADM formulation of classical (2+1)-dimensional gravity. As we saw in chapter 4, however, the classical theory can be described equally well in terms of geometric structures and the holonomies of flat connections. The two classical descriptions are ultimately equivalent, but they are quite different in spirit: the ADM formalism depicts a spatial geometry evolving in time, while the geometric structure formalism views the entire spacetime as a single 'timeless' entity.

The corresponding quantum theories are just as different. In particular, while ADM quantization incorporates a clearly defined time variable, the quantum theory of geometric structures, which we shall develop in this chapter, will be a 'quantum gravity without time' [170, 171]. Nevertheless, the two quantum theories, like their classical counterparts, are closely related: the quantum theory of geometric structures will turn out to be a sort of 'Heisenberg picture' that complements the 'Schrödinger picture' of ADM quantization.

The approach of this chapter is commonly called the connection representation, and closely resembles the (3+1)-dimensional connection representation developed by Ashtekar *et al.* [11]. The name comes from the fact that the basic variables – in this case, the geometric structures of chapter 4 – are associated with the spin connection rather than the metric. In particular, the 'configuration space' of geometric structures is the space of $SO(2,1)$ holonomies of the spin connection.

6.1 Covariant phase space

Our starting point for this chapter is the classical description of (2+1)-dimensional gravity developed in chapter 4. There, we obtained a geometric description of the *solutions* of the field equations. Our first question is therefore what it means to quantize a space of solutions. The general answer is given by the program of covariant canonical quantization [15, 85, 276].

Covariant canonical quantization begins with the the simple but profound observation that the phase space of a well-behaved classical theory is isomorphic to the space of classical solutions. Specifically, let $\mathscr{C}$ be an arbitrary (but fixed) Cauchy surface. Then a point in the phase space determines initial data on $\mathscr{C}$, which in turn determine a unique solution. Conversely, a classical solution restricted to $\mathscr{C}$ determines a point in the phase space. This observation suggests a solution to the old 'covariant vs. canonical' debate in quantum gravity, by offering a manifestly covariant approach to canonical quantization.

We next need a symplectic structure – a set of Poisson brackets – on the space of solutions [175, 275]. Suppose our classical theory is defined by the action[*]

$$I = \int_M \mathbf{L}[\phi], \qquad (6.1)$$

where M is an n-dimensional manifold, the Lagrangian $\mathbf{L}$ is an n-form on M, and ϕ denotes a collection of fields that take their values in some other manifold N. Fields are thus mappings $\phi : M \to N$, and the action is a functional on the space $\mathscr{F}$ of such mappings. We can think of an infinitesimal variation δ as an exterior derivative on $\mathscr{F}$; thus, for example, $\delta \mathbf{L}$ is an n-form on M and a one-form of $\mathscr{F}$.

Let us write the equations of motion in the general form $\mathbf{E} = 0$. A variation of $\mathbf{L}$ is then

$$\delta \mathbf{L} = \mathbf{E}\delta\phi + d\Theta, \qquad (6.2)$$

where the n-form $d\Theta[\phi, \delta\phi]$ is a total derivative whose integral gives the boundary terms in the variation. We define the symplectic current $(n-1)$-form to be

$$\Omega[\phi, \delta_1\phi, \delta_2\phi] = \delta_1 \left(\Theta[\phi, \delta_2\phi] \right) - \delta_2 \left(\Theta[\phi, \delta_1\phi] \right). \qquad (6.3)$$

Suppose we now restrict our attention to the space $\mathscr{F}_0$ of solutions of the field equations $\mathbf{E} = 0$, and to variations $\delta\phi$ within this space of solutions. Restricted to $\mathscr{F}_0$, the integral of $\Omega[\phi, \delta_1\phi, \delta_2\phi]$ over a Cauchy surface $\mathscr{C}$,

$$\widehat{\Omega}[\phi, \delta_1\phi, \delta_2\phi] = \int_{\mathscr{C}} \Omega[\phi, \delta_1\phi, \delta_2\phi], \qquad (6.4)$$

---

[*] In this section, following Wald [275], I denote differential forms on M by bold face letters.

is independent of the choice of $\mathscr{C}$, since

$$\int_{\mathscr{C}_1} \Omega[\phi, \delta_1\phi, \delta_2\phi] - \int_{\mathscr{C}_2} \Omega[\phi, \delta_1\phi, \delta_2\phi] = \int_{M_{12}} d\Omega[\phi, \delta_1\phi, \delta_2\phi]$$

$$= \delta_1 (\delta_2 \mathbf{L} - \mathbf{E}\delta_2\phi) - \delta_2 (\delta_1 \mathbf{L} - \mathbf{E}\delta_1\phi) = 0, \quad (6.5)$$

where M_{12} denotes the region of M between $\mathscr{C}_1$ and $\mathscr{C}_2$ and I have used equation (6.2).

The quantity $\widehat{\Omega}[\phi, \delta_1\phi, \delta_2\phi]$ is a two-form on $\mathscr{F}_0$, and defines a symplectic form on the space of solutions. Under the isomorphism between the space of solutions and the standard phase space, $\widehat{\Omega}$ may be shown to map to the usual phase space symplectic form. When the theory defined by $\mathbf{L}$ has gauge invariances, $\widehat{\Omega}$ is really only a presymplectic form – it is degenerate in the 'gauge directions' – but it projects to a genuine symplectic form on the space of solutions modulo gauge transformations. Many of the tools of ordinary Hamiltonian theory, such as Noether's theorem for the construction of conserved charges, can be translated into this covariant phase space formalism [175, 275].

For (2+1)-dimensional gravity, we can take $\mathbf{L}$ to be the first order Lagrangian of chapter 2,

$$\mathbf{L} = -2\left\{ e^a \wedge \left(d\omega_a + \frac{1}{2}\epsilon_{abc}\omega^b \wedge \omega^c \right) + \frac{\Lambda}{6}\epsilon_{abc}e^a \wedge e^b \wedge e^c \right\}. \quad (6.6)$$

It is then easy to see that

$$\Theta[\phi, \delta\phi] = 2 (e^a \wedge \delta\omega_a), \quad (6.7)$$

and hence

$$\widehat{\Omega}[e, \omega, \delta_1 e, \delta_2\omega] = 2 \int_M (\delta_1 e^a \wedge \delta_2\omega_a). \quad (6.8)$$

This last expression is very similar to the Poisson bracket structure of equation (2.100), but it is now restricted to gauge-invariant parametrizations of classical solutions of the equations of motion. As a consequence, it defines a symplectic structure on the space of geometric structures $\widetilde{\mathscr{M}}$ of chapter 4.

Now recall that when the cosmological constant is zero, $\widetilde{\mathscr{M}}$ is a cotangent bundle, $\widetilde{\mathscr{M}} = T^*\widetilde{\mathscr{N}}$. As such, it has a natural symplectic structure. This symplectic structure is almost, but not quite, the same as the physical symplectic structure obtained from $\widehat{\Omega}$.

Consider, for example, the solutions (3.76)–(3.78) for spacetimes with the topology $[0, 1] \times T^2$ with $\Lambda = 0$:

$$
\begin{aligned}
e^0 &= e^{-t}dt & \omega^0 &= 0 \\
e^1 &= adx + bdy & \omega^1 &= -(\lambda dx + \mu dy) \\
e^2 &= e^{-t}(\lambda dx + \mu dy) & \omega^2 &= 0.
\end{aligned}
\tag{6.9}
$$

Substituting into (6.8), we see that

$$
\widehat{\Omega}[\delta e, \delta \omega] = 2(\delta \mu \wedge \delta a - \delta \lambda \wedge \delta b),
\tag{6.10}
$$

in agreement with equation (3.82). Hence λ and b are conjugate variables, as are μ and a. The cotangent bundle structure of $T^*\mathcal{N}$, on the other hand, comes from viewing e^a as a variation of ω^a. Comparing e^a and ω^a in (6.9), we see this structure would make λ conjugate to a – the term adx in e^1 is obtained by varying λdx in ω^1 – and μ conjugate to b. To obtain the physical symplectic form $\widehat{\Omega}$, we must 'twist' this naive structure, essentially interchanging x and y.

We can understand the need for this twist by recalling that two holonomies $\rho[\gamma_1]$ and $\rho[\gamma_2]$ have nontrivial Poisson brackets only if γ_1 and γ_2 are intersecting, nonhomotopic curves. The naive cotangent bundle structure, on the other hand, associates a translational holonomy along a curve γ with an $SO(2, 1)$ holonomy along the *same* curve. The physical symplectic form $\widehat{\Omega}$ reflects both the cotangent bundle structure of $T^*\widetilde{\mathcal{N}}$ and the symplectic structure of the space of homotopy classes discussed in section 7 of chapter 4.

While this 'twisting' can cause some complications, the physical symplectic form $\widehat{\Omega}$ can always be computed from equation (6.8), provided we know how to parametrize e^a and ω^a in terms of holonomies. Given a set of coordinates μ_α on the $SO(2, 1)$ moduli space $\widetilde{\mathcal{N}}$ and a parametrization $\omega = \omega(\mu_\alpha)$, equation (6.8) allows us to define a set of conjugate momenta

$$
\pi^\alpha = 2 \int_\Sigma e^a \wedge \left(\frac{\partial \omega_a}{\partial \mu_\alpha} \right).
\tag{6.11}
$$

(Compare equation (2.33) in the ADM formalism.) Note that the first-order action is now simply

$$
I = \int dt\, \pi^\alpha \dot{\mu}_\alpha,
\tag{6.12}
$$

with no Hamiltonian. This is not surprising – the holonomies we have chosen as our fundamental variables are, after all, constants of motion. But it is also not surprising that this time independence will complicate the interpretation of the quantum theory.

6.2 Quantizing geometric structures

With this preparatory work completed, it is now almost trivial to quantize the space of geometric structures. The covariant phase space $\mathscr{M}$ is parametrized by coordinates μ_α and π^α, which are canonically conjugate in the classical theory. Moreover, despite the complications described above, the π^α label cotangent vectors to the $SO(2,1)$ moduli space $\widetilde{\mathscr{N}}$. To quantize, we simply trade Poisson brackets for commutators,

$$[\hat{\mu}_\alpha, \hat{\pi}^\beta] = i\delta_\alpha^\beta, \tag{6.13}$$

and allow operators to act on a Hilbert space $\mathscr{H}_{conn}$ of square-integrable functions of the μ_α.

For the torus, for example, the nonvanishing commutators are

$$[\hat{\mu}, \hat{a}] = [\hat{b}, \hat{\lambda}] = -\frac{i}{2}. \tag{6.14}$$

We can take our wave functions to be functions $\psi(\lambda, \mu)$, and represent $\hat{a}$ and $\hat{b}$ as

$$\hat{a} = \frac{i}{2}\frac{\partial}{\partial\mu}, \qquad \hat{b} = -\frac{i}{2}\frac{\partial}{\partial\lambda}. \tag{6.15}$$

This is, in a sense, all there is to the quantum theory. There is no Hamiltonian, but this reflects the classical choice of time-independent holonomies as our fundamental observables. Given a wave function, we can determine the spectrum and the expectation values of the holonomies, and we can ask for the probability that the universe has a given geometric structure; that is, we can obtain probabilistic answers for all of the basic questions we asked in chapter 4.

Of course, without a Hamiltonian we seem to be missing some rather important dynamical information. The recovery of this information will be discussed in the next section. First, however, an important technical issue must be clarified.

Following the procedure of the last chapter, we should presumably require that our wave functions be invariant under the action of the mapping class group. For the torus, this group is generated by the transformations

$$\begin{aligned}
S &: (\lambda,\mu) \to (\mu,-\lambda) & (a,b) &\to (b,-a) \\
T &: (\lambda,\mu) \to (\lambda,\mu+\lambda) & (a,b) &\to (a,b+a),
\end{aligned} \tag{6.16}$$

as is most easily seen by considering the action of Dehn twists on the fundamental group $\pi_1(T^2)$. Now, however, we encounter a rather serious problem. In the ADM formalism, the action (3.73) of the modular group on Teichmüller space was properly discontinuous,[†] and the moduli space $\mathcal{N} = \widetilde{\mathcal{N}} / \sim$ obtained from Teichmüller space by factoring out this action was well-behaved. For spatial topologies of genus $g > 1$, this is true as well for the action of the mapping class group on the configuration space of $SO(2,1)$ holonomies. For the torus, however, the action (6.16) is *not* properly discontinuous, and in fact the quotient of $\widetilde{\mathcal{N}}$ by this group is nowhere locally a manifold [219]. This means, among other things, that there are no continuous, nonconstant modular invariant wave functions $\psi(\lambda, \mu)$. Worse yet, Peldan has shown that there are no continuous, nonconstant wave functions that transform under any finite-dimensional representation of the modular group, although the possibility of projective representations has not yet been investigated. We may try dropping the requirement of continuity – quantum mechanics really only requires measurable wave functions – but even then, severe difficulties occur in the attempt to implement modular invariance [128].

There are several possible reactions to this dilemma. One is to choose a different polarization, that is, a different splitting of the phase space $\widetilde{\mathcal{M}}$ into 'positions' and 'momenta'. To see that this can affect the outcome, consider the (admittedly not very natural) choice of configuration space variables [220]

$$(x, y) = \left(\frac{ab + \lambda\mu}{a^2 + \lambda^2}, \frac{a\mu - b\lambda}{a^2 + \lambda^2} \right). \tag{6.17}$$

It is easy to check that these coordinates transform among themselves under (6.16), and the transformations are in fact properly discontinuous. This is a trick, of course: x and y as defined here are just the ADM moduli τ_1 and τ_2 of equation (3.81) at time $t = 0$, and we already know that these moduli transform nicely. But the trick illustrates the fact that problems with the mapping class group may be avoidable.

In particular, it is not known whether the holonomy variables $r_a^{\pm}$ of equation (4.33), the natural variables for (2+1)-dimensional gravity with a negative cosmological constant, lead to a quantum theory that is well-behaved with respect to the modular group. These variables, like the (λ, μ, a, b), have a simple symplectic structure that can be obtained from (4.39):

$$\widehat{\Omega}[\delta e, \delta \omega] = \ell(\delta r_1^+ \wedge \delta r_2^+ - \delta r_1^- \wedge \delta r_2^-). \tag{6.18}$$

The corresponding quantum theory has wave functions $\psi(r_2^+, r_2^-)$, with operators

$$\hat{r}_1^+ = -\frac{i}{\ell} \frac{\partial}{\partial r_2^+}, \qquad \hat{r}_1^- = \frac{i}{\ell} \frac{\partial}{\partial r_2^-} \tag{6.19}$$

[†] See appendix A for a brief introduction to the concept of a discontinuous group action.

analogous to (6.15). The action of the mapping class group on the $\hat{r}_a^{\pm}$ can be shown to be

$$S : (\hat{r}_1^{\pm}, \hat{r}_2^{\pm}) \rightarrow (\hat{r}_2^{\pm}, -\hat{r}_1^{\pm})$$
$$T : (\hat{r}_1^{\pm}, \hat{r}_2^{\pm}) \rightarrow (\hat{r}_1^{\pm} + \hat{r}_2^{\pm}, \hat{r}_2^{\pm}). \tag{6.20}$$

In contrast to the case of a vanishing cosmological constant, the modular group now mixes positions and momenta. The generators of modular transformations can be represented as operators

$$\hat{S} = \exp\left\{ \frac{\pi i \ell}{4} \left[(\hat{r}_1^+)^2 + (\hat{r}_2^+)^2 - (\hat{r}_1^-)^2 - (\hat{r}_2^-)^2 \right] \right\}$$
$$\hat{T} = \exp\left\{ \frac{i\ell}{2} \left[(\hat{r}_2^+)^2 - (\hat{r}_2^-)^2 \right] \right\}, \tag{6.21}$$

and the requirement of invariance becomes a set of differential equations. Using the properties of $SL(2, \mathbf{R})$ representations, it has recently been shown that some invariant wave functions exist, but the known solutions are nonnormalizable, and the general characteristics of invariant wave functions are not yet understood [70].

Another reaction to the problem of imposing modular invariance is to simply drop the requirement. This might seem too drastic a step: the modular transformations are, after all, diffeomorphisms, and we could reasonably insist that any theory of quantum gravity behave well under diffeomorphisms. But we shall see below that even without an explicit demand of modular invariance, the relationship between the connection representation and ADM quantization can teach us a good deal about the modular group.

6.3 Relating quantizations

Quite apart from the technical difficulties with the modular group, the quantum theory constructed above exhibits a fundamental conceptual problem: it has no visible dynamics. We have built an archetypical 'frozen time' theory, in which the observables are constants of motion and the evolution is completely hidden. This feature is not peculiar to 2+1 dimensions, but is a generic problem in quantum gravity. The natural candidate for the Hamiltonian in quantum gravity is the Hamiltonian constraint $\mathcal{H}$ of chapter 2, but at least for a closed universe, we have seen that $\mathcal{H}$ is zero on physical states. We can attempt to restore dynamics by introducing an explicit time-slicing, as in chapter 5, but as we have seen, such a procedure violates at least the spirit of general covariance.

This 'timeless' nature of quantum gravity already exists in the classical formalism: the description of (2+1)-dimensional gravity in terms of geometric structures exhibits the same lack of explicit dynamics. In the classical theory, however, this problem causes no great concern. Given a set of holonomies, we can determine a classical spacetime, and given that spacetime, we can answer any dynamical question we care to ask: for example, how do the moduli of a slice of constant TrK evolve with time? One of the principle achievements of (2+1)-dimensional quantum gravity has been to show that this procedure can be carried over to a quantum theory.

As usual, we shall concentrate on the topology $[0, 1] \times T^2$. Suppose first that the cosmological constant Λ vanishes. In equation (4.31), we constructed a function $\tau(T; \lambda, \mu, a, b)$ whose physical meaning was 'the value of the modulus on the slice of constant mean curvature $TrK = -T$ in a universe with a geometric structure characterized by the holonomies λ, μ, a, and b', along with similar functions $p(T; \lambda, \mu, a, b)$ and $H_{red}(T; \lambda, \mu, a, b)$. We can now turn these functions into operators acting on the Hilbert space $\mathcal{H}_{conn}$ of the connection representation [52, 68, 69]. If we choose the orderings

$$\hat{\tau}(T) = \left(\hat{a} + \frac{i\hat{\lambda}}{T} \right)^{-1} \left(\hat{b} + \frac{i\hat{\mu}}{T} \right)$$

$$\hat{p}(T) = -iT \left(\hat{a} - \frac{i\hat{\lambda}}{T} \right)^2$$

$$\hat{H}_{red}(T) = \frac{\hat{\mu}\hat{a} - \hat{b}\hat{\lambda}}{T}, \tag{6.22}$$

then the modular transformations (6.16) induce the standard modular transformations of $\hat{\tau}$ and $\hat{p}$. Moreover, it may be checked that the commutation relations (6.13) in the connection representation induce the ADM commutation relations (5.3), and that the operators $\hat{\tau}$ and $\hat{p}$ satisfy the 'Heisenberg equations of motion'

$$i\frac{d\hat{\tau}}{dT} = [\hat{\tau}, \hat{H}_{red}], \qquad i\frac{d\hat{p}}{dT} = [\hat{p}, \hat{H}_{red}]. \tag{6.23}$$

The ADM moduli and momenta may thus be viewed as 'time'-dependent operators in the connection representation, obeying standard quantum mechanical equations of motion. Given a state $\psi(\lambda, \mu)$, we can thus determine, for example, the probability that the slice with a given value of TrK has a certain modulus.

Operators such as $\hat{\tau}(T)$ are not exactly functions of time: after all, we have completely solved the constraints, and there is no time coordinate left in the connection representation. Indeed, for each value of the parameter T, $\hat{\tau}(T)$ and $\hat{p}(T)$ are constants of motion. These operators can best be understood as examples of Rovelli's 'evolving constants of motion', operators that are time-independent but nevertheless provide information about dynamics [230, 231]. We shall return to the question of how to interpret such operators at the end of this chapter.

We can learn more about these operators – and in the process, gain some insight into the problem of modular invariance – by diagonalizing $\hat{\tau}(T)$ and its adjoint to obtain 'Schrödinger picture' wave functions [53]. A long but straightforward computation shows that the function

$$K(T;\tau,\bar{\tau};\lambda,\mu) = \frac{\mu - \tau\lambda}{\pi\tau_2^{1/2}T} \exp\left\{-\frac{i}{\tau_2 T}|\mu - \tau\lambda|^2\right\} \tag{6.24}$$

is a simultaneous eigenfunction of the operators $\hat{\tau}$ and $\hat{\tau}^\dagger$ of equation (6.22), with

$$\hat{\tau}K(T;\tau,\bar{\tau};\lambda,\mu) = \tau K(T;\tau,\bar{\tau};\lambda,\mu),$$
$$\hat{\tau}^\dagger K(T;\tau,\bar{\tau};\lambda,\mu) = \bar{\tau}K(T;\tau,\bar{\tau};\lambda,\mu). \tag{6.25}$$

The μ and λ dependence of K is determined by the eigenvalue equations, while the prefactor is fixed by the normalization

$$\int d\lambda d\mu \bar{K}(T;\tau',\bar{\tau}';\lambda,\mu)K(T;\tau,\bar{\tau};\lambda,\mu) = \tau_2^2\delta^2(\tau - \tau'). \tag{6.26}$$

The factor of τ_2^2 on the right hand side of this expression converts the ordinary delta function into a delta function with respect to the inner product (5.34). Note that the integration in (6.26) is over the entire λ-μ plane, not over some fundamental region for the modular group.

The function $K(T;\tau,\bar{\tau};\lambda,\mu)$ can be viewed as the transformation matrix between the connection representation and the ADM representation of the preceding chapter. It has the following important properties:

1. Under the modular transformations (3.10) of τ, K transforms as

$$S : K(T;\tau,\bar{\tau};\lambda,\mu) \rightarrow K(T;-\frac{1}{\tau},-\frac{1}{\bar{\tau}};\lambda,\mu)$$
$$= \left(\frac{\tau}{\bar{\tau}}\right)^{-1/2} K(T;\tau,\bar{\tau};-\mu,\lambda)$$
$$T : K(T;\tau,\bar{\tau};\lambda,\mu) \rightarrow K(T;\tau+1,\bar{\tau}+1;\lambda,\mu)$$
$$= K(T;\tau,\bar{\tau};\lambda,\mu-\lambda). \tag{6.27}$$

Comparing equation (5.29), we can identify these transformations as those of an automorphic form of weight $-1/2$, combined with the inverse of the transformations (6.16) of λ and μ.

2. The connection representation Hamiltonian (6.22) acts on K as

$$\hat{H}_{red} K(T;\tau,\bar{\tau};\lambda,\mu) = -i\frac{\partial}{\partial T} K(T;\tau,\bar{\tau};\lambda,\mu). \qquad (6.28)$$

Moreover, the weight $-1/2$ ADM Hamiltonian of chapter 5 acts as

$$\Delta_{-1/2} K(T;\tau,\bar{\tau};\lambda,\mu) = \left(iT\frac{\partial}{\partial T}\right)^2 K(T;\tau,\bar{\tau};\lambda,\mu), \qquad (6.29)$$

which is the square of the Schrödinger equation (5.37).

We can now write an arbitrary connection representation wave function $\psi(\lambda,\mu)$ as a superposition

$$\psi(\lambda,\mu) = \int_{\mathscr{F}} \frac{d^2\tau}{\tau_2{}^2} K(T;\tau,\bar{\tau};\lambda,\mu)\bar{\tilde{\psi}}(\tau,\bar{\tau},T) = \langle \tilde{\psi}, K \rangle, \qquad (6.30)$$

where property 1 suggests that we should require $\tilde{\psi}$ to be an automorphic form of weight $-1/2$. The coefficient $\tilde{\psi}$, which we shall interpret as a 'Schrödinger picture' wave function, has been written as a function of T, but the T dependence is not arbitrary, since the left-hand side of equation (6.30) is independent of T. Indeed, we can use equation (6.26) to solve for $\tilde{\psi}(\tau,\bar{\tau},T)$; differentiating with respect to T and using (6.29), we find that

$$\left(iT\frac{\partial}{\partial T}\right)^2 \tilde{\psi}(\tau,\bar{\tau},T) = \Delta_{-1/2}\tilde{\psi}(\tau,\bar{\tau},T), \qquad (6.31)$$

which is the square of the Schrödinger equation (5.37) for an automorphic form of weight $-1/2$.

To determine the region of integration $\mathscr{F}$ in (6.30), let us consider the normalization of wave functions:

$$\langle \psi | \psi \rangle = \qquad (6.32)$$

$$\int d\lambda d\mu \int_{\mathscr{F}} \frac{d^2\tau}{\tau_2{}^2} \frac{d^2\tau'}{\tau_2'^2} \bar{K}(T;\tau,\bar{\tau};\lambda,\mu) K(T;\tau',\bar{\tau}';\lambda,\mu)\tilde{\psi}(\tau,\bar{\tau},T)\bar{\tilde{\psi}}(\tau',\bar{\tau}',T)$$

$$= \int_{\mathscr{F}} \frac{d^2\tau}{\tau_2{}^2} \frac{d^2\tau'}{\tau_2'^2} \tau_2{}^2 \delta^2(\tau-\tau')\tilde{\psi}(\tau,\bar{\tau},T)\bar{\tilde{\psi}}(\tau',\bar{\tau}',T) = \int_{\mathscr{F}} \frac{d^2\tau}{\tau_2{}^2} |\tilde{\psi}(\tau,\bar{\tau},T)|^2,$$

where I have used the orthogonality relations (6.26) for K. But $\tilde{\psi}$ is an automorphic form, so this last integral will be normalized if $\mathscr{F}$ is a fundamental region for action of the modular group on Teichmüller space, such as the 'keyhole region' described in appendix A.

We can now return to the question of how the modular group acts in the connection representation. A wave function defined by equation (6.30) is not modular invariant. On the contrary, if g is an arbitrary element of the modular group, $\psi(\lambda, \mu)$ and $\psi(g\lambda, g\mu)$ are orthogonal. For example,

$$\langle \psi | S \psi \rangle =$$

$$\int d\lambda d\mu \int_{\mathscr{F}} \frac{d^2\tau}{\tau_2{}^2} \frac{d^2\tau'}{\tau_2'^2} \bar{K}(T; \tau, \bar{\tau}; \lambda, \mu) K(T; \tau', \bar{\tau}'; \mu, -\lambda) \tilde{\psi}(\tau, \bar{\tau}, T) \bar{\tilde{\psi}}(\tau', \bar{\tau}', T)$$

$$= \int_{\mathscr{F}} \frac{d^2\tau}{\tau_2{}^2} \frac{d^2\tau'}{\tau_2'^2} \tau_2{}^2 \delta^2(\tau - S\tau') \left(\frac{\tau'}{\bar{\tau}'} \right)^{1/2} \tilde{\psi}(\tau, \bar{\tau}, T) \bar{\tilde{\psi}}(\tau', \bar{\tau}', T), \qquad (6.33)$$

where I have used the transformation properties (6.27) of K. But S maps τ' from $\mathscr{F}$ to an adjacent fundamental region, and except at isolated points, fundamental regions for the mapping class group are disjoint, so

$$\int_{\mathscr{F}} \frac{d^2\tau}{\tau_2{}^2} \tau_2{}^2 \delta^2(\tau - S\tau') = 0, \qquad (6.34)$$

and the inner product (6.33) thus vanishes. The same argument holds if S is replaced by T, or by any nontrivial combination of S and T, that is, by any element of the modular group.

A similar argument shows that if $\psi(\lambda, \mu)$ and $\chi(\lambda, \mu)$ are *any* two wave functions defined by (6.30), with integrals over the same fundamental region $\mathscr{F}$, then

$$\langle \psi | g \chi \rangle = 0 \qquad (6.35)$$

for any element g of the modular group. Viewed as an integral transform, the mapping (6.30) thus takes the weight $-1/2$ ADM Hilbert space $\mathscr{F}^{-1/2}$ of chapter 5 into an infinite number of orthogonal subspaces of the Hilbert space $\mathscr{H}_{conn}$, which are interchanged by elements of the modular group. A version of modular invariance for the connection representation would thus be the restriction of $\mathscr{H}_{conn}$ to one of these subspaces.

As an automorphic form of weight $-1/2$, $\tilde{\psi}(\tau', \bar{\tau}', T)$ is essentially a spinor on moduli space. It was shown in reference [53] that the second-order Klein–Gordon-like equation (6.31) may be converted to a first-order 'Dirac equation' for a two-component spinor

$$\Psi = \left(\begin{array}{c} \tilde{\psi} \\ \tilde{\pi} \end{array} \right) \qquad (6.36)$$

on moduli space, where the lower component is essentially a time-reversed ADM wave function. The appearance of such time-reversed solutions should not be surprising. Classically, a solution of the Einstein equations with the topology $[0, 1] \times T^2$ can describe either an expanding or a collapsing universe. As we saw in the preceding chapter, the Schrödinger equation in the ADM formalism requires a choice of sign in the definition of the square root $\Delta_n^{1/2}$, and that choice amounts to a restriction to either purely expansion or pure collapse. In the connection representation, the same ambiguity has reappeared in a slightly different guise.

Are weight $-1/2$ automorphic forms really necessary in the connection representation? The first indication of their importance came from the transformation properties (6.27) of the eigenfunctions of $\hat{\tau}$ and $\hat{\tau}^\dagger$. But the definition (6.22) of these moduli operators involves an implicit choice of operator ordering, which is by no means unique. It may be shown that different, though perhaps less natural, choices of ordering can lead to automorphic forms of arbitrary weight [55]. Equivalently, at least in this simple context, the choice of inner product in the Hilbert space $\mathcal{H}_{conn}$ is ambiguous, and different choices lead to different adjoints $\hat{\tau}^\dagger$. For example, if one chooses the inner product

$$\langle \psi | \chi \rangle = \int d\lambda d\mu (\hat{p}^{1/2} \psi)^* (\hat{p}^{1/2} \chi), \tag{6.37}$$

where $\hat{p}$ is the momentum operator of equation (6.22), it may be shown that the corresponding ADM wave functions are weight zero forms, that is, ordinary automorphic functions. By finding a series of canonical transformations that link the ADM representation and the connection representation, Anderson has found a similar, although not identical, measure [7]. The connection representation has not resolved the ambiguity described in the preceding chapter, but has merely transformed it to an ambiguity in the choice of operator ordering.

So far, we have interpreted the function $K(T; \tau, \bar{\tau}; \lambda, \mu)$ as a transformation matrix between the connection representation and the ADM representation, as manifested by the transformation (6.30). But this function has another interpretation as well: for fixed values of τ, $\bar{\tau}$, and T, it may be viewed as the state $|\tau, \bar{\tau}, T\rangle$ of a torus universe with modulus τ at time T, expressed in the connection representation. This dual role is familiar from ordinary quantum mechanics, where the function e^{ipx} is both the transformation matrix between the position and the momentum representations and the position representation wave function for the state $|p\rangle$. As in ordinary quantum mechanics, the matrix element

$$\langle \tau', \bar{\tau}', T' | \tau, \bar{\tau}, T \rangle = \int d\lambda d\mu \bar{K}(T'; \tau', \bar{\tau}'; \lambda, \mu) K(T; \tau, \bar{\tau}; \lambda, \mu) \tag{6.38}$$

thus gives an amplitude for a spatial geometry determined by the modulus τ at time T to evolve to a geometry determined by the modulus τ' at time T'. This amplitude has been worked out explicitly by Ezawa, who shows that it is peaked at precisely the classical trajectory (3.70) [107]. The quantum theory thus approximates the classical behavior, as it should.

A similar set of constructions is possible when $\Lambda < 0$. Starting with equation (4.38), we can define operators

$$\hat{\tau}(\bar{t}) = \left(\hat{r}_2^- e^{i\bar{t}/\ell} + \hat{r}_2^+ e^{-i\bar{t}/\ell}\right)^{-1} \left(\hat{r}_1^- e^{i\bar{t}/\ell} + \hat{r}_1^+ e^{-i\bar{t}/\ell}\right)$$

$$\hat{p}(\bar{t}) = \frac{i\ell}{2\sin(2\bar{t}/\ell)} \left(\hat{r}_2^+ e^{i\bar{t}/\ell} + \hat{r}_2^- e^{-i\bar{t}/\ell}\right)^2$$

$$\hat{H}_{red}(\bar{t}) = \frac{\ell^2}{4} \sin\frac{2\bar{t}}{\ell} \left(\hat{r}_1^- \hat{r}_2^+ - \hat{r}_1^+ \hat{r}_2^-\right). \tag{6.39}$$

Again, it may be checked that the modular transformations (6.16) induce the standard transformations of $\hat{\tau}$ and $\hat{p}$, that the connection representation commutators (6.13) induce the correct ADM commutators, and that $\hat{\tau}$ and $\hat{p}$ satisfy the appropriate Heisenberg equations of motion. If we take our states in the connection representation to be functions of r_2^+ and r_2^-, the transformation matrix analogous to (6.24) is then

$$K(\bar{t}; \tau, \bar{\tau}; r_2^+, r_2^-) = \frac{\ell\tau_2}{2\pi} \left(\sin\frac{2\bar{t}}{\ell}\right)^{-1/2} \left(r_2^+ e^{i\bar{t}/\ell} + r_2^- e^{-i\bar{t}/\ell}\right) \times \tag{6.40}$$

$$\exp\frac{i\ell}{2} \left\{\tau_1 \left[(r_2^+)^2 - (r_2^-)^2\right] - \tau_2 \left[(r_2^+)^2 + 2r_2^+ r_2^- \sec\frac{2\bar{t}}{\ell} + (r_2^-)^2\right] \cot\frac{2\bar{t}}{\ell}\right\}.$$

A rather long computation now shows that the transformation properties (6.27) again hold, with S and T now represented by the operators of equation (6.21). Moreover, the analog of (6.29) continues to be valid:

$$\Delta_{-1/2} K(\bar{t}; \tau, \bar{\tau}; r_2^+, r_2^-) = \left(i\frac{\ell}{2} \sin\frac{2\bar{t}}{\ell}\frac{\partial}{\partial\bar{t}}\right)^2 K(\bar{t}; \tau, \bar{\tau}; r_2^+, r_2^-), \tag{6.41}$$

which by (4.40) is again the square of the Schrödinger equation of the preceding chapter [70].

6.4 Ashtekar variables

The connection representation of the preceding sections was based on a particular set of classical solutions, those with nondegenerate metrics. For the torus, the classical phase space was even more restricted: we excluded those geometries on $[0, 1] \times T^2$, discussed briefly in chapter 3, for which the T^2 slices were not spacelike. By loosening these restrictions, we can obtain new quantum theories that are not quite the same as those we have seen so far. In particular, the Ashtekar approach to (2+1)-dimensional quantum gravity relaxes the requirement of nondegeneracy of the metric, thus enlarging the classical phase space and the quantum Hilbert space.

In the connection representation developed in the last two sections, the phase space $T^*\mathcal{N}$ was a space of flat metrics and spin connections. This flatness originates in the constraints of chapter 2,

$$\mathscr{C}^a[\omega, e] = \frac{1}{2}\epsilon^{ij}D_i e_j = \frac{1}{2}\epsilon^{ij}\left[\partial_i e_j{}^a - \partial_j e_i{}^a + \epsilon^{abc}(\omega_{ib}e_{jc} - \omega_{jb}e_{ic})\right] = 0, \tag{6.42}$$

$$\widetilde{\mathscr{C}}^a[\omega, e] = \frac{1}{2}\epsilon^{ij}F_{ij} = \frac{1}{2}\epsilon^{ij}\left[\partial_i\omega_j{}^a - \partial_j\omega_i{}^a + \epsilon^{abc}\omega_{ib}\omega_{jc}\right] = 0, \tag{6.43}$$

where D_i denotes the $SO(2,1)$ gauge-covariant derivative. The Ashtekar constraints, on the other hand, take a slightly different form. The phase space variables are again a spin connection and a triad, or more commonly a spin connection and a densitized triad,[‡]

$$\widetilde{E}^{ia} = \epsilon^{ij}e_j{}^a, \tag{6.44}$$

which clearly carries exactly the same information as e. The constraints are now written in a form identical to that of the Ashtekar constraints in 3+1 dimensions [11, 35],

$$D_i\widetilde{E}^{ia} = 0, \tag{6.45}$$

$$\widetilde{E}^i{}_a F_{ij}^a = 0, \tag{6.46}$$

$$\epsilon_{abc}\widetilde{E}^{ib}\widetilde{E}^{jc}F_{ij}^a = 0. \tag{6.47}$$

The constraint (6.45) is clearly equivalent to (6.42). When the spatial metric $g_{ij} = e_i{}^a e_{ja}$ is nondegenerate, it is not hard to show that the constraints (6.46)–(6.47) are equivalent to (6.43). Indeed, using the definition of $\widetilde{E}^{ia}$ and the fact that $F_{ij}^a = \epsilon_{ij}\widetilde{\mathscr{C}}^a$, we can rewrite these constraints as

$$e_{ia}\widetilde{\mathscr{C}}^a = 0$$

$$\epsilon_{abc}\epsilon^{kl}e_k{}^b e_l{}^c\widetilde{\mathscr{C}}^a = 0 \tag{6.48}$$

and the equivalence follows from some elementary algebra.

The Ashtekar constraints (6.45)–(6.47) admit additional solutions, however, for which the spatial metric is degenerate and the curvature F_{ij}^a does not vanish. The full space of solutions is not completely understood, but portions have been investigated. Barbero and Varadarajan have argued that the full phase space has an infinite number of degrees of freedom [25]. Let us define a 'flat patch' of a surface Σ as a region in which the constraints (6.42)–(6.43) are satisfied, and a 'null patch' as a region in which the metric is degenerate, the constraints (6.45)–(6.47) are satisfied,

---

[‡] It is traditional in papers using Ashtekar variables to use a tilde over a letter to denote a density of weight one.

and curvature satisfies $\tilde{\mathscr{C}}^a\tilde{\mathscr{C}}_a = 0$. Then if initial data are chosen so that Σ consists of a collection of disjoint flat patches separated by null patches, this pattern is preserved by evolution, and the holonomies in each of the flat patches become independent observables. The consequences for the quantum theory are not yet understood, although some work has been done in the loop representation and in minisuperspace models [26, 109]. At a minimum, the (2+1)-dimensional model serves as a warning that the choice of how to treat degenerate metrics in quantum gravity can drastically affect the outcome.

6.5 More pros and cons

The quantum theory developed in this chapter has managed to avoid some of the conceptual difficulties of the reduced phase space ADM quantization of chapter 5. In particular, we have not had to choose any particular time slicing, and have thus preserved general covariance. This manifest covariance comes at a price: the physical interpretation of the resulting 'frozen time' formalism, and in particular the identification of dynamical behavior, becomes much less obvious. But we have seen in at least one example that the problems of interpretation can be overcome, and that dynamical questions can be answered through the construction of appropriate operators.

Operators of the sort we have studied here are not quite 'dynamical'. The modulus $\hat{\tau}(T)$, for example, should not be thought of as a single, time-dependent operator. Rather, for each value of the parameter T, $\hat{\tau}(T)$ is defined throughout spacetime, and commutes with the constraints – it is a constant of motion, or in Kuchař's terms, a 'perennial' [172]. We must interpret $\hat{\tau}(T)$ as a one-parameter family of operators, indexed by a number T, that are defined everywhere.

Nevertheless, we can use the correspondence principle and our knowledge of the classical theory to give $\hat{\tau}(T)$ its dynamical interpretation as the operator that represents the modulus of the slice $TrK = -T$. The rather mysterious role of time may become clearer when we compare the quantum theory to the classical theory of chapter 4. There, too, the spacetime geometry was described completely in terms of constants of motion, and no explicit time coordinate was ever introduced. Cosgrove has called the resulting quantum theory 'quantization after evolution': the dynamics is completely solved classically and reexpressed in terms of correlations among time-independent constants of motion, which can then be studied quantum mechanically [81]. This point of view is closely related to Rovelli's description of quantum mechanics in terms of 'evolving constants of motion' [230, 231].

We can now return to a question raised in the preceding chapter: are the quantum theories corresponding to different choices of time-slicing equivalent? In this chapter, we considered only a particular set of 'dynamical' operators, those corresponding to the York time-slicing $TrK = -T$. But nothing prevents us from looking at a different choice of slicing. Given any classical choice of time, with slices labeled by some parameter t, we can construct the corresponding classical observables $\tau(t)$ and $p(t)$ and convert them into operators in the connection representation. Just as we did for the York slicing, we can then diagonalize $\tau(t)$ and its adjoint and transform to a 'Schrödinger picture'.

By construction, though, this 'Schrödinger picture' is equivalent to the original 'Heisenberg picture' of the connection representation. In particular, a transition amplitude between slices t_i and t_f depends only on the initial operator $\hat\tau(t_i)$ and its adjoint, the final operator $\hat\tau(t_f)$ and its adjoint, and a time-independent wave function $\psi(\lambda, \mu)$. Provided that the operator orderings in $\hat\tau(t_i)$, $\hat\tau(t_f)$, and their adjoints are fixed, the choice of intermediate slicing cannot affect the amplitude.

This last proviso, however, emphasizes a remaining arbitrariness in the connection representation. Dynamical quantities like $\tau(T)$ are classically well-determined, but the process of constructing operators is fraught with ordering ambiguities. In particular, the 'natural' ordering with respect to one choice of time-slicing may not be the same as the 'natural' ordering with respect to another. This problem reflects a genuine physical uncertainty: when we measure a classical gravitational field, we do not know exactly which quantum operators we are observing. This problem is already present in ordinary quantum mechanics, although it is usually ignored because we 'know' the right variables. But we know only because of the results of experiments, and these are sorely lacking in quantum gravity.

A similar ambiguity arises in the choice of polarization, that is, the splitting of the phase space into 'positions' and 'momenta'. We saw briefly in section 2 that this choice can, at a minimum, affect the representation of the modular group. There is no proof that different choices of polarization lead to equivalent quantum theories, and there is no obvious way to choose a 'correct' polarization, although the mapping class group might offer some guidance.

There is another difficulty with the connection representation, which is in some sense 'merely technical' but which has the potential to cause serious problems. To construct the dynamical operators needed for a physical interpretation, we must know the classical solutions of the equations of motion; more specifically, we need a parametrization of the classical solutions in terms of a set of constants of motion. Equation (6.22) for the operator $\hat\tau$, for example, was obtained from the classical equation

(4.31), which in turn depended on our explicit knowledge of the general solution of the vacuum field equations. For (3+1)-dimensional gravity, this is clearly impractical; even for (2+1)-dimensional gravity on a manifold with a spatial topology of genus $g > 1$, the general solution for the metric is not known explicitly in terms of holonomy variables. The obvious approach to this problem is to try to develop a perturbative formalism for operators, based on a perturbative approach to the classical theory. So far, however, such a 'perturbative covariant canonical quantization' has not been investigated.

7
Operator algebras and loops

In its most basic form, quantization is the process of replacing the Poisson brackets of a classical theory with commutators of operators acting on some Hilbert space,

$$\{x, y\} \rightarrow \frac{1}{i\hbar}[\hat{x}, \hat{y}].$$ (7.1)

Many of the ambiguities in quantization arise because this is not, strictly speaking, possible: even for the simplest classical theories, factor-ordering difficulties prevent us from simultaneously making the substitutions (7.1) for all classical variables x and y [263]. The usual procedure is therefore to pick out a set $\mathscr{S}$ of 'elementary classical variables', which are to have unambiguous quantum analogs, and to construct the remaining operators from elements of $\mathscr{S}$. The set $\mathscr{S}$ must be small enough to close under Poisson brackets, and it must be large enough that the remaining classical variables can be constructed from sums and products of its elements, or as limits of such sums and products. But apart from these basic requirements, there is no known 'right' way to choose the elementary classical variables.

In chapter 5, we took as our elementary variables the moduli of the spatial metric at $TrK = -T$, along with their conjugate momenta. In chapter 6, we chose a set of coordinates μ_α on the space of $SO(2, 1)$ holonomies, along with their conjugate momenta. But there is another obvious candidate for $\mathscr{S}$: the $ISO(2, 1)$ holonomies themselves, or a suitable subset of these holonomies. As we saw in chapter 4, the Poisson algebra of the holonomies will lead to commutators that are considerably more complicated that the simple canonical brackets (5.3) or (6.13), but we can nevertheless look for an operator representation.

The literature contains two fundamental approaches to the quantization of the algebra of holonomies. The first, developed by Nelson and Regge, is most easily implemented for the case of a negative cosmological constant

$\Lambda = -1/\ell^2$; the group of holonomies is then $SL(2, \mathbf{R}) \times SL(2, \mathbf{R})$ [205, 206, 207, 208, 209]. The second, initiated by Ashtekar and his collaborators, starts with the case $\Lambda = 0$, and invokes a transformation from the holonomy representation to a dual 'loop representation' [10, 11, 13].

7.1 The operator algebra of Nelson and Regge

The approach to (2+1)-dimensional quantum gravity initiated by Nelson and Regge is based on the algebra of holonomies described in chapter 4. The program can be summarized as follows:

1. Choose a sufficiently small but complete (or overcomplete) set $\mathscr{S}$ of holonomies $R_\alpha = \frac{1}{2} Tr \rho[\gamma_\alpha]$, whose classical algebra closes under Poisson brackets (4.73). If the set is overcomplete, find a complete set of classical relations $F_\beta = 0$.

2. Represent the holonomies as operators $\hat{R}_\alpha \in \hat{\mathscr{S}}$, with a commutator algebra obtained, up to possible higher-order terms in $\hbar$, by the substitution (7.1), and with the symmetric ordering

$$xy \rightarrow \frac{1}{2} (\hat{x}\hat{y} + \hat{y}\hat{x}) \tag{7.2}$$

for products of holonomies occurring in the commutator algebra. The set $\hat{\mathscr{S}}$ must remain closed under this algebra.

3. If the set $\mathscr{S}$ is overcomplete, find operator orderings for the relations such that

$$\left[\hat{F}_\beta, \hat{R}_\alpha\right] = 0 \tag{7.3}$$

for all holonomies in $\hat{\mathscr{S}}$.

4. Represent the operator algebra as an algebra acting on some Hilbert space of states, restricted to the subspace of states annihilated by the $\hat{F}_\beta$.

5. Find the action of the mapping class group on this space of states, and restrict to invariant or suitably covariant states.

As usual, this program is simplest for spacetimes with the topology $[0, 1] \times T^2$, and the steps 1–5 have been largely completed for the case of a negative cosmological constant. For our elementary set of holonomies, we take the set $\{R_1^\pm, R_2^\pm, R_{12}^\pm\}$ described in section 7 of chapter 4, with the relations $F^\pm = 0$. The operator algebra corresponding to the brackets

(4.74) can then be determined by direct substitution of operators for classical variables; one finds that

$$\hat{R}_1^{\pm}\hat{R}_2^{\pm}e^{\pm i\theta} - \hat{R}_2^{\pm}\hat{R}_1^{\pm}e^{\mp i\theta} = \pm 2i\sin\theta\,\hat{R}_{12}^{\pm} \quad \text{and cyclical permutations}$$
(7.4)

with

$$\tan\theta = -\frac{2\pi G}{\ell},$$
(7.5)

while the relators $F^{\pm}$ become

$$\hat{F}^{\pm} = 1 - \tan^2\theta - e^{\pm 2i\theta}\left((\hat{R}_1^{\pm})^2 + (\hat{R}_{12}^{\pm})^2\right)$$
$$- e^{\mp 2i\theta}(\hat{R}_2^{\pm})^2 + 2e^{\pm i\theta}\cos\theta\,\hat{R}_1^{\pm}\hat{R}_2^{\pm}\hat{R}_{12}^{\pm}.$$
(7.6)

It is straightforward to check that the $\hat{F}^{\pm}$ do, in fact, commute with the $\hat{R}_{\alpha}$.

This algebra is not a Lie algebra, but it is related to the Lie algebra of the quantum group $U_q(sl(2)) \times U_q(sl(2))$, where $q = \exp 4i\theta$ and the $\hat{F}^{\pm}$ are related to the cyclically invariant q-Casimir. The commutation relations (7.4) are invariant under the modular transformations

$$S : \hat{R}_1^{\pm} \to \hat{R}_2^{\pm}, \quad \hat{R}_2^{\pm} \to \hat{R}_1^{\pm}, \quad \hat{R}_{12}^{\pm} \to \hat{R}_1^{\pm}\hat{R}_2^{\pm} + \hat{R}_2^{\pm}\hat{R}_1^{\pm} - \hat{R}_{12}^{\pm}$$
$$T : \hat{R}_1^{\pm} \to \hat{R}_{12}^{\pm}, \quad \hat{R}_2^{\pm} \to \hat{R}_2^{\pm}, \quad \hat{R}_{12}^{\pm} \to \hat{R}_{12}^{\pm}\hat{R}_2^{\pm} + \hat{R}_2^{\pm}\hat{R}_{12}^{\pm} - \hat{R}_1^{\pm},$$
(7.7)

which reduce to the standard transformations in the classical limit.

To obtain a quantum theory, we must now represent this algebra as an algebra of operators acting on a suitable Hilbert space. To obtain an explicit representation, let us define

$$\hat{R}_1^{\pm} = \sec\theta\cosh\frac{\hat{r}_1^{\pm}}{2}$$

$$\hat{R}_2^{\pm} = \sec\theta\cosh\frac{\hat{r}_2^{\pm}}{2}$$

$$\hat{R}_{12}^{\pm} = \sec\theta\cosh\frac{\hat{r}_1^{\pm} + \hat{r}_2^{\pm}}{2}.$$
(7.8)

It is then easily checked that the relations (7.4) hold provided

$$[\hat{r}_1^{\pm}, \hat{r}_2^{\pm}] = \pm 8i\theta, \qquad [\hat{r}_{\alpha}^{+}, \hat{r}_{\beta}^{-}] = 0,$$
(7.9)

and that the relators $\hat{F}^{\pm}$ are identically zero.

For small θ, that is, small cosmological constant, the commutators (7.9) reduce to those found in the connection representation of chapter 6, up to terms of order θ^3. Moreover, for any value of θ, equation (7.9) can be interpreted as a set of connection representation commutators appropriately rescaled. Hence the results of chapter 6 can be carried over directly to the Nelson–Regge quantization. In particular, the representations discussed in chapter 6 automatically give rise to representations of the operator algebra (7.4). We have reduced the Nelson–Regge approach to the connection representation, but with a small adjustment: the coupling constant has been slightly deformed.

The results of chapter 6 also give us a physical interpretation of the holonomies $R_\alpha^\pm$. For instance, we can use the holonomies to construct operators representing the ADM modulus: as in equations (6.39) and (4.40), we can define

$$\hat{\tau}(\bar{t}) = \left(\hat{r}_2^- e^{i\bar{t}/\ell} + \hat{r}_2^+ e^{-i\bar{t}/\ell}\right)^{-1}\left(\hat{r}_1^- e^{i\bar{t}/\ell} + \hat{r}_1^+ e^{-i\bar{t}/\ell}\right)$$

$$\hat{p}(\bar{t}) = \frac{i\ell}{2\sin(2\bar{t}/\ell)}\left(\hat{r}_2^+ e^{i\bar{t}/\ell} + \hat{r}_2^- e^{-i\bar{t}/\ell}\right)^2$$

$$\hat{H}_{red}(\bar{t}) = \csc\frac{2\bar{t}}{\ell}\left(\hat{r}_1^- \hat{r}_2^+ - \hat{r}_1^+ \hat{r}_2^-\right). \tag{7.10}$$

Equation (7.9) then gives the commutators

$$[\hat{\tau}^\dagger, \hat{p}] = [\hat{\tau}, \hat{p}^\dagger] = -16i\ell\theta, \qquad [\hat{\tau}, \hat{p}] = [\hat{\tau}^\dagger, \hat{p}^\dagger] = 0 \tag{7.11}$$

and the Heisenberg equations of motion

$$[\hat{p}, \hat{H}_{red}] = -8i\ell\theta\frac{d\hat{p}}{dt}, \qquad [\hat{\tau}, \hat{H}_{red}] = -8i\ell\theta\frac{d\hat{\tau}}{dt}, \tag{7.12}$$

which differ from the corresponding equations in ADM quantization by terms of order $O(\hbar^3)$, small when the cosmological constant Λ is small.

An important advantage of this algebraic approach is that it may be generalizable to more complicated spatial topologies. For a surface Σ of genus $g > 1$, consider the set of closed curves t_i, $i = 1, 2, \ldots, 2g+2$, shown in figure 7.1. These curves can be constructed explicitly in terms of the standard generators A_i and B_i of $\pi_1(\Sigma)$ described in appendix A, and they satisfy the three relations

$$t_1 t_3 \ldots t_{2g+1} = 1$$
$$t_2 t_4 \ldots t_{2g+2} = 1$$
$$t_1 t_2 t_3 \ldots t_{2g+2} = 1. \tag{7.13}$$

If we cut Σ along the t_i, the surface splits into four pieces, each a $(2g+2)$-sided polygon, as shown in figure 7.2 for the case of genus 2.

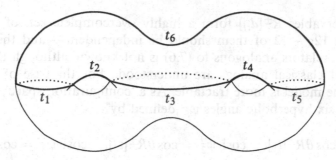

Fig. 7.1. A genus two surface can be cut along the curves $t_1, \ldots, t_6$, shown here as dark lines, to form four hexagons as shown in figure 7.2. (Only the top halves of t_2, t_4, and t_6 are shown in this figure.)

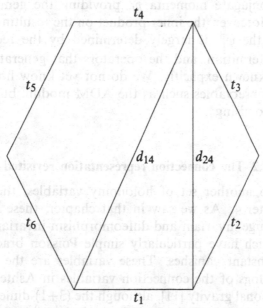

Fig. 7.2. Four hexagons of this type are formed by cutting open the surface in figure 7.1. The curves t_i, d_{14}, and d_{24} are eight of the fifteen whose holonomy algebra has been studied by Nelson and Regge.

Nelson and Regge consider the collection of $(g + 1)(2g + 1)$ curves $d_{ij} = t_i t_{i+1} \ldots t_{j-1}$, which can be represented as line segments connecting vertices i and j of a polygon such as that of figure 7.2. Any two curves in this set intersect at most twice, and the classical Poisson algebra (4.73) of holonomies $R^{\pm}[d_{ij}]$ closes. Moreover, the quantum algebra generalizing (7.4) may again be shown to close, providing a starting point for the quantum theory.

The observables $R^\pm[d_{ij}]$ form a highly overcomplete set, of course – in all, only $12g - 12$ of them should be independent – and the general form of the relators analogous to (7.6) is not known, although the rather complicated classical analogs are understood. For the case of genus 2, however, the model is more tractable. As a 'configuration space', one may choose the six hyperbolic angles $\psi_i^\pm$ defined by[*]

$$\cosh \psi_1^\pm = \cos \theta R^\pm[t_1], \quad \cosh \psi_2^\pm = \cos \theta R^\pm[t_3], \quad \cosh \psi_3^\pm = \cos \theta R^\pm[t_5].$$
$$(7.14)$$

It is evident from figure 7.2 that the curves t_1, t_3, and t_5 do not intersect, so the $\psi_i^\pm$ commute with one another. Nelson and Regge have shown explicitly how to reconstruct the remaining operators $\hat{R}^\pm[d_{ij}]$ from the $\psi_i^\pm$ and their conjugate momenta p_i, providing the genus 2 analog of equation (7.8). Moreover, the inner product on the resulting Hilbert space of functions of the $\psi_i^\pm$ is largely determined by the requirement that the $\hat{R}^\pm[d_{ij}]$ be Hermitian, and the operators that generate the mapping class group are known explicitly. We do not yet know how to construct time-dependent observables such as the ADM moduli, but the approach appears to be promising.

7.2 The connection representation revisited

We turn next to another set of holonomy variables, the functions T^0 and T^1 of chapter 4. As we saw in that chapter, these provide a convenient set of gauge-invariant and diffeomorphism-invariant functions on phase space, which have particularly simple Poisson brackets when the cosmological constant vanishes. These variables are the natural (2+1)-dimensional analogs of the connection variables in Ashtekar's approach to (3+1)-dimensional gravity [11], although the (3+1)-dimensional version is rather more complicated: in 3+1 dimensions, T^0 and T^1 depend on loops, not just on homotopy classes of loops, and they no longer commute with the constraints.

Our goal is to promote the algebra (4.84)–(4.86) of Poisson brackets to an operator algebra, with operators $\hat{T}^0[\gamma]$ and $\hat{T}^1[\gamma]$ acting on a suitable Hilbert space. One approach to this problem is fairly simple, at least in principle, but will turn out not give us anything new. As usual, we consider a spacetime with the topology $[0, 1] \times \Sigma$, with Σ a closed surface. We would like to find a representation – technically, a $\star$-representation –

---

[*] Conventions for the angle θ, here given by (7.5), differ among papers in this field.

of the abstract algebra

$$\left[\widehat{T}^0[\gamma_1], \widehat{T}^0[\gamma_2]\right] = 0 \tag{7.15}$$

$$\left[\widehat{T}^1[\gamma_1], \widehat{T}^0[\gamma_2]\right] = -\frac{i}{8} \sum_i \epsilon(p_i; \gamma_1, \gamma_2) \left(\widehat{T}^0[\gamma_1 \cdot_i \gamma_2] - \widehat{T}^0[\gamma_1 \cdot_i \gamma_2^{-1}]\right)$$

$$\left[\widehat{T}^1[\gamma_1], \widehat{T}^1[\gamma_2]\right] = -\frac{i}{8} \sum_i \epsilon(p_i; \gamma_1, \gamma_2) \left(\widehat{T}^1[\gamma_1 \cdot_i \gamma_2] - \widehat{T}^1[\gamma_1 \cdot_i \gamma_2^{-1}]\right)$$

with $(\widehat{T}^0)^\star = \widehat{T}^0$ and $(\widehat{T}^1)^\star = \widehat{T}^1$, subject to the relations

$$\widehat{T}^0[\gamma_1]\widehat{T}^0[\gamma_2] = \frac{1}{2}\left(\widehat{T}^0[\gamma_1 \cdot \gamma_2] + \widehat{T}^0[\gamma_1 \cdot \gamma_2^{-1}]\right)$$

$$\widehat{T}^0[\gamma_1]\widehat{T}^1[\gamma_2] + \widehat{T}^0[\gamma_2]\widehat{T}^1[\gamma_1] = \frac{1}{2}\left(\widehat{T}^1[\gamma_1 \cdot \gamma_2] + \widehat{T}^1[\gamma_1 \cdot \gamma_2^{-1}]\right) \tag{7.16}$$

and

$$\widehat{T}^0[0] = 1, \qquad \widehat{T}^0[\gamma] = \widehat{T}^0[\gamma^{-1}], \qquad \widehat{T}^0[\gamma_1 \cdot \gamma_2] = \widehat{T}^0[\gamma_2 \cdot \gamma_1]$$

$$\widehat{T}^1[0] = 0, \qquad \widehat{T}^1[\gamma] = \widehat{T}^1[\gamma^{-1}], \qquad \widehat{T}^1[\gamma_1 \cdot \gamma_2] = \widehat{T}^1[\gamma_2 \cdot \gamma_1]. \tag{7.17}$$

Note first that the functions T^0 depend on the spin connection alone, and are therefore defined on the space of equivalence classes of flat $SO(2,1)$ connections, that is, the moduli space $\widetilde{\mathcal{N}}$ of hyperbolic structures on Σ. Indeed, by equation (4.81), T^0 is just the trace of the homomorphism

$$\rho_0 : \pi_1(\Sigma) \to SO(2,1) \tag{7.18}$$

of chapter 4, and the trace is invariant under the equivalence relation in (4.15). Just as in chapter 6, we can therefore choose our Hilbert space $\mathcal{H}_{conn}$ to be a suitable space of functions on $\widetilde{\mathcal{N}}$ – or a space of half-densities if we wish to avoid the problem of choosing a measure – and we can represent the $T^0[\gamma]$ as acting by multiplication.

Moreover, when viewed as a function of the triad and spin connection, the variable $T^1[\omega, e]$ can be regarded as a cotangent vector to $\widetilde{\mathcal{N}}$ at the point representing ω, with a direction determined by e. Recall that if ω is a flat connection and e satisfies the Einstein field equations (2.62), then to first order in t, $\omega + te$ is also a flat connection. Equation (4.82) can thus be rewritten as

$$T^1[\omega, e] = 2 \frac{d}{dt} T^0[\omega + te] \Big|_{t=0} \tag{7.19}$$

clearly defining a cotangent vector to $\widetilde{\mathcal{N}}$. As in section 1 of chapter 6, the physical symplectic structure is not quite the natural symplectic structure of the cotangent bundle $T^* \widetilde{\mathcal{N}}$, but the difference is calculable, and once again allows us to represent $T^1[\omega, e]$ as a derivative acting on $\mathcal{H}_{conn}$.

The result of this quantization is not new: it is precisely the connection representation of chapter 6. For the torus, for instance, homotopy classes of curves are labeled by two winding numbers m and n, and T^0 and T^1 can be computed explicitly from the triad and spin connection derived in chapter 3: by equation (4.87),

$$\widehat{T}^0[m, n] = \cosh \frac{m\hat{\lambda} + n\hat{\mu}}{2}, \qquad \widehat{T}^1[m, n] = -(m\hat{a} + n\hat{b}) \sinh \frac{m\hat{\lambda} + n\hat{\mu}}{2}. \tag{7.20}$$

If we now impose the connection representation commutators

$$[\hat{\mu}, \hat{a}] = -[\hat{\lambda}, \hat{b}] = -\frac{i}{2}, \tag{7.21}$$

of chapter 6, it is easily checked that the $\widehat{T}^0$ and $\widehat{T}^1$ commutation relations (7.15) are satisfied. Conversely, given a Hilbert space upon which the $\widehat{T}^0[m, n]$ act by multiplication, the commutators (7.15) may be used to derive the connection representation commutators. Note that the identities (7.16) and (7.17) are also automatically satisfied, since they simply express properties of $SL(2, \mathbf{R})$ traces.

7.3 The loop representation

There is another way to look at the operator algebra of $\widehat{T}^0[\gamma]$ and $\widehat{T}^1[\gamma]$ which leads to an important new approach to quantum gravity, the loop representation. So far, we have been thinking of the operators $\widehat{T}$ as a set of functions of the triad and spin connection, indexed by loops γ. Our wave functions are thus functionals of ω, or, more precisely, functions on the moduli space $\mathcal{N}$ of flat spin connections. This dependence is clear in the expressions (7.20) for $\widehat{T}^0$ and $\widehat{T}^1$ on the torus. However, we could equally well have chosen to view the $\widehat{T}$ operators as functions of loops – or in 2+1 dimensions, of homotopy classes of loops – indexed by e and ω. Wave functions would then most naturally be functions of loops or sets of loops. This change of viewpoint is rather like the decision in ordinary quantum mechanics to view a wave function e^{ipq} as a function on momentum space, indexed by q, rather than a function on position space, indexed by p. This analogy will be strengthened below, when we derive a 'loop transform' as a sort of generalized Fourier transform.

The starting point for the loop representation is again the algebra (7.15) and the identities (7.16)–(7.17). We would now like to find a representation of this algebra that makes no explicit reference to the triad or spin connection. There are a number of equivalent approaches to this problem, but perhaps the simplest is due to Ashtekar [11].[†] We begin by introducing a vacuum state $|0\rangle$, defined by the condition that

$$\hat{T}^1[\gamma]|0\rangle = 0 \qquad (7.22)$$

for all homotopy classes $[\gamma]$. We can act on this state by arbitrary products of $\hat{T}^0$ and $\hat{T}^1$. By using the algebra (7.15)–(7.16), however, we can move any factor $\hat{T}^1$ to the left, where it will annihilate the vacuum state. Similarly, any product $\hat{T}^0[\gamma_1]\hat{T}^0[\gamma_2]$ can be reduced by (7.15)–(7.16) to a sum of single copies of $\hat{T}^0$. By repeated application of these rearrangements, we can finally express any state as a linear combination

$$|\Psi\rangle = \sum c_i|[\gamma_i]\rangle, \qquad \text{with } |[\gamma_i]\rangle = \hat{T}^0[\gamma_i]|0\rangle. \qquad (7.23)$$

The use of $[\gamma_i]$ to label the state $|[\gamma_i]\rangle$ is suggestive, but also slightly misleading: not all distinct homotopy classes give rise to different states. For example, by (7.17), $|[\alpha]\rangle = |[\alpha^{-1}]\rangle$ and $|[\alpha \cdot \beta]\rangle = |[\beta \cdot \alpha]\rangle$. Ashtekar has suggested the term 'equitopy' to describe the equivalence relation

$$[\gamma_1] \sim [\gamma_2] \qquad \text{if } T^0[\gamma_1](\omega) = T^0[\gamma_2](\omega) \text{ for all flat spin connections } \omega. \qquad (7.24)$$

We should think of the states $|[\gamma_i]\rangle$ as being labeled by equitopy classes of curves.

Even with this caveat, the set of states $|[\gamma_i]\rangle$ is overcomplete. For example, by (7.16) and (7.17),

$$|[\alpha \cdot \beta \cdot \gamma]\rangle + |[\alpha \cdot \gamma^{-1} \cdot \beta^{-1}]\rangle - |[\alpha \cdot \gamma \cdot \beta]\rangle - |[\alpha \cdot \beta^{-1} \cdot \gamma^{-1}]\rangle = 0. \qquad (7.25)$$

Identities of this type determine a subspace defined by the relation

$$\sum c_i|[\gamma_i]\rangle = 0 \qquad \text{whenever } \sum c_i T^0[\gamma_i] = 0. \qquad (7.26)$$

We can now define the Hilbert space for the loop representation as the quotient space $\mathcal{H}_{loop} = \mathcal{H}'/\sim$, where $\mathcal{H}'$ is the space generated by the $|[\gamma_i]\rangle$ and $\sim$ is the equivalence relation (7.26). The inner product on $\mathcal{H}_{loop}$ can be defined by demanding that the $|[\gamma_i]\rangle$ be orthonormal. The actions of $\hat{T}^0$ and $\hat{T}^1$ on $\mathcal{H}_{loop}$ follow from the definition of the basis states and the algebra (7.15)–(7.17):

$$\hat{T}^0[\alpha]|[\beta]\rangle = \frac{1}{2}\left(|[\alpha] \cdot [\beta]\rangle + |[\alpha] \cdot [\beta]^{-1}\rangle\right)$$

$$\hat{T}^1[\alpha]|[\beta]\rangle = -\frac{i}{8}\sum_i \epsilon(p_i; \alpha, \beta)\left(|[\alpha \cdot_i \beta]\rangle - |[\alpha \cdot_i \beta^{-1}]\rangle\right). \qquad (7.27)$$

[†] This approach is similar to that of DePietri and Rovelli in 3+1 dimensions [86].

To determine the physical meaning of the loop representation, it is useful to again specialize to the torus, and to try to relate loop states to the connection representation states of chapter 6. The resulting relationship is known as the loop transform. As we shall see, a formal transformation between representations is easy to write down, but there are mathematical subtleties that make the result rather difficult to interpret.

Let us assume that our spacetime has the topology $[0,1] \times T^2$. As in the preceding section, homotopy classes of curves on T^2 may be labeled by integers m and n. The basis states in the loop representation can thus be denoted $|m,n\rangle$, with the relation $|m,n\rangle \sim |-m,-n\rangle$ coming from (7.17). Equation (7.27) then becomes

$$\hat{T}^0[m,n]|p,q\rangle = \frac{1}{2}\left(|m+p,n+q\rangle + |m-p,n-q\rangle\right)$$
$$\hat{T}^1[m,n]|p,q\rangle = -\frac{i}{8}(mq-np)\left(|m+p,n+q\rangle - |m-p,n-q\rangle\right).$$
$$(7.28)$$

The $\hat{T}^0[m,n]$ commute, so we can look for simultaneous eigenstates, that is, linear combinations

$$|\Psi\rangle = \sum_{p,q} c_{pq}|p,q\rangle \tag{7.29}$$

such that

$$\hat{T}^0[m,n]|\Psi\rangle = T^0[m,n]|\Psi\rangle \tag{7.30}$$

for some numbers $T^0[m,n]$. Equation (7.28) gives a set of recursion relations for the c_{pq}, and it may be verified that they are satisfied by

$$|\lambda_0,\mu_0\rangle = \sum_{p,q} \cosh\left(\frac{p\lambda_0 + q\mu_0}{2}\right)|p,q\rangle,$$
$$\hat{T}^0[m,n]|\lambda_0,\mu_0\rangle = \cosh\left(\frac{m\lambda_0 + n\mu_0}{2}\right)|\lambda_0,\mu_0\rangle \tag{7.31}$$

where λ_0 and μ_0 are arbitrary real numbers. Comparing equation (7.20), we see that the state $|\lambda_0,\mu_0\rangle$ is an eigenstate of the operators $\hat{\lambda}$ and $\hat{\mu}$ of the connection representation with eigenvalues λ_0 and μ_0. Moreover, it is easy to verify that the action of the operator $\hat{T}^1[m,n]$ on such a state also agrees with the connection representation action determined by (7.20).

A general connection representation state can now be written as a superposition

$$\psi(\lambda,\mu) = \langle\psi|\lambda,\mu\rangle \tag{7.32}$$
$$= \sum_{p,q} \cosh\left(\frac{p\lambda + q\mu}{2}\right)\langle\psi|p,q\rangle = \sum_{p,q}\cosh\left(\frac{p\lambda+q\mu}{2}\right)\tilde{\psi}(p,q),$$

or equivalently

$$\psi(\lambda, \mu) = \sum_{[\gamma]} T^0[\gamma](\lambda, \mu)\tilde{\psi}([\gamma]). \tag{7.33}$$

This relation is the inverse loop transform, relating loop and connection representation states. If the hyperbolic cosines in equation (7.31) were instead cosines, we would have an ordinary Fourier transform, closely analogous to the transformation between the position and momentum representations in quantum mechanics. We could then easily invert (7.31) to express $\tilde{\psi}([\gamma])$ in terms of connection wave functions. There are, in fact, solutions to the eigenvalue problem (7.30) for which such a procedure is natural: if the hyperbolic cosines in equation (7.31) are replaced by cosines, the equation remains valid. The resulting states have eigenvalues $T^0[m, n] \leq 1$, however, and correspond to the geometries with non-spacelike slices discussed briefly in chapters 3 and 4. For the physically interesting sector, no such obvious inverse exists.

It is nevertheless interesting to write the 'inverse' transform

$$\tilde{\psi}'(p, q) = \int d\lambda d\mu \cosh\left(\frac{p\lambda + q\mu}{2}\right)\psi(\lambda, \mu), \tag{7.34}$$

that is,

$$\tilde{\psi}'[\gamma] = \int d\lambda d\mu\, T^0[\gamma](\lambda, \mu)\psi(\lambda, \mu). \tag{7.35}$$

This integral is the (2+1)-dimensional version of the (3+1)-dimensional loop transform in the Ashtekar approach to quantum gravity. We can determine the action of a $\hat{T}$ operator on a state $\tilde{\psi}'(p, q)$ by starting with the corresponding action on $\psi(\lambda, \mu)$ in the connection representation, as given by equation (7.20). Using standard identities for products of hyperbolic functions, we find that

$$\hat{T}^0[m, n]\tilde{\psi}'(p, q) = \int d\lambda d\mu \cosh\left(\frac{p\lambda + q\mu}{2}\right)\cosh\left(\frac{m\lambda + n\mu}{2}\right)\psi(\lambda, \mu)$$

$$= \frac{1}{2}\left(\tilde{\psi}'(p + m, q + n) + \tilde{\psi}'(p - m, q - n)\right),$$

$$\hat{T}^1[m, n]\tilde{\psi}'(p, q)$$

$$= -\frac{i}{2}\int d\lambda d\mu \cosh\left(\frac{p\lambda + q\mu}{2}\right)\sinh\left(\frac{m\lambda + n\mu}{2}\right)\left[m\frac{\partial}{\partial\mu} - n\frac{\partial}{\partial\lambda}\right]\psi(\lambda, \mu)$$

$$= -\frac{i}{4}(np - mq)\int d\lambda d\mu \sinh\left(\frac{p\lambda + q\mu}{2}\right)\sinh\left(\frac{m\lambda + n\mu}{2}\right)\psi(\lambda, \mu)$$

$$= -\frac{i}{8}(np - mq)\left(\tilde{\psi}'(p + m, q + n) - \tilde{\psi}'(p - m, q - n)\right), \tag{7.36}$$

in agreement with equation (7.28). The transform (7.34) thus gives a good formal representation of the action of the loop operators, and could be used as a starting point for the definition of the loop representation. It is not, however, the inverse of the expansion (7.31); indeed, the insertion of (7.31) into (7.34) leads to a divergent integral.

In fact, the transformation (7.34) is generally rather poorly behaved. Let $\mathcal{H}_{conn}$ denote the Hilbert space in the connection representation, as defined in chapter 6, and view (7.34) as a mapping from $\mathcal{H}_{conn}$ to $\mathcal{H}_{loop}$. As Marolf first showed, there is a dense subspace of $\mathcal{H}_{conn}$ that lies in the kernel of this mapping; that is, there is a dense set of connection representation states which transform to zero [185]. There is, however, another dense subspace $\mathcal{V} \subset \mathcal{H}_{conn}$ on which the transformation (7.34) is faithful, and this subspace is preserved by $\widehat{T}^0$ and $\widehat{T}^1$. It is possible to use such a subspace to make the loop transform (7.34) well-defined: one can define the transform on $\mathcal{V}$, determine the inner product and the action of the $\widehat{T}$ operators on the resulting loop states, and then form the Cauchy completion of this space to define $\mathcal{H}_{loop}$. The subspace $\mathcal{V}$ is not unique, but the result of this procedure is independent of the choice of $\mathcal{V}$, in the sense that all of the resulting loop representations are isomorphic. However, many of the states in the Cauchy completion are no longer functions of loops in any clear sense.

Alternatively, Ashtekar and Loll have showed that the loop transform (7.34) can be made into an isomorphism by choosing a more elaborate measure for the integral over the moduli μ and λ, selected to make various integrals converge [14]. This change induces order $\hbar$ corrections to the action (7.28) of the $\widehat{T}^1$ operators, and the new measure must be chosen carefully to ensure that the added terms can be written in terms of $\widehat{T}^0$ operators. Such a choice is possible. Unfortunately, the inner products between loop states become considerably more complex, as does the action of the mapping class group.

These two approaches to the loop transform – starting with a dense subspace $\mathcal{V}$ or choosing a more complicated measure – are unitarily equivalent. In both, however, the physical interpretation of the quantum theory becomes obscure. Nor is the role of the mapping class group understood: we saw in chapter 6 that the space $\mathcal{H}_{conn}$ splits into infinitely many orthogonal subspaces that transform among themselves, but the corresponding splitting of $\mathcal{H}_{loop}$ has not been studied.

As with the Nelson–Regge variables, there has been some work on extending the loop representation to spacetimes with more complicated spatial topologies. Starting with the set of independent loop variables described at the end of chapter 4, the Ashtekar–Loll approach has been applied to spacetime topologies $[0, 1] \times \Sigma$ with Σ a genus $g > 1$ surface, although many of the details remain to be worked out [178, 179]. Again, however, the physical interpretation of the loop representation in terms of clearly defined geometrical observables remains difficult.

An interesting exception to this shortage of good observables is the length operator $\hat{\ell}[C]$. Classically, the length of a curve C on Σ is

$$
\ell[C] = \int_C ds \left(g_{ij} \frac{dC^i}{ds} \frac{dC^j}{ds} \right)^{1/2}
$$

$$
= \int_C ds \left(e_i{}^a \frac{dC^i}{ds} e_j{}^b \frac{dC^j}{ds} \eta_{ab} \right)^{1/2}. \tag{7.37}
$$

In the quantum theory, the product of triads in (7.37) is ill-defined, and must be regularized, for instance by point-splitting. To carry out this procedure, let γ_x^ϵ be a loop of radius ϵ centered at x, and define

$$
T_{ij}(s,t) = 2\,Tr\left[\rho_0(t,s) e_i{}^a(\gamma_x^\epsilon(s)) \mathscr{J}_a \rho_0(s,t) e_j{}^b(\gamma_x^\epsilon(t)) \mathscr{J}_b \right]. \tag{7.38}
$$

Here $\rho_0(s,t)$ is the holonomy of chapter 4 along the segment of γ_x^ϵ between s and t,

$$
\rho_0(s,t) = P\exp\left\{ \int_s^t du\; \omega_i{}^a \frac{dx^i(\gamma(u))}{du} \mathscr{J}_a \right\}, \tag{7.39}
$$

and the $\mathscr{J}^a$ are generators of $SU(2)$ in the spin $1/2$ representation. This definition of T_{ij} is clearly related to the definition (4.82) of $T^1[\gamma]$, and T_{ij} has an easily computed set of Poisson brackets with the T^0 and T^1 variables that parametrize Σ. On the other hand, it is clear that as ϵ becomes small, $T_{ij}(0,\pi)$ approaches $g_{ij}(x)$. We can therefore define a length operator in the loop representation by

$$
\hat{\ell}[C] = \lim_{\epsilon \to 0} \int_C ds \left(\hat{T}_{ij}(0,\pi) \frac{dC^i}{ds} \frac{dC^j}{ds} \right)^{1/2}. \tag{7.40}
$$

Rovelli has shown that this operator has a discrete spectrum, with eigenvalues that are multiples of $L_{Pl}/2$ [232].

Now, $\hat{\ell}[C]$ is not the *unique* quantum gravitational length operator: it is, after all, determined from the classical expression (7.37) only up to terms of order $\hbar$. Nor is $\hat{\ell}[C]$ really a physical operator, since for a generic curve C, $\hat{\ell}[C]$ does not commute with the constraints. The source of this difficulty is easy to understand from equation (7.37): the constraints act on the metric g_{ij} as generators of diffeomorphisms, but if C is defined in a field-independent manner, the constraints commute with the derivatives dC^i/ds rather than transforming them. Nevertheless, it is certainly intriguing that a natural choice of quantization yields lengths that are quantized in units of half the Planck length, and it is plausible – although not proven – that this quantization of the length spectrum will continue to hold for physical length operators acting on geometrically defined curves. We shall return briefly to this question in chapter 11 when we discuss the Turaev–Viro lattice model.

8

The Wheeler–DeWitt equation

The approaches to quantization described in chapters 5–7, although quite different, share one common feature. They are all 'reduced phase space' quantizations, quantum theories based on the true physical degrees of freedom of the classical theory.

As we saw in chapter 2, not all of the degrees of freedom that determine the metric in general relativity have physical significance; many are 'pure gauge', describing coordinate choices rather than dynamics, and can be eliminated by solving the constraints and factoring out the diffeomorphisms. Indeed, we have seen that in 2+1 dimensions only a finite number of the '$6 \times \infty^3$' metric degrees of freedom are physical. In each of the preceding approaches to quantization, our first step was to eliminate the nonphysical degrees of freedom, sometimes explicitly and sometimes indirectly through a clever choice of variables; only then were the remaining degrees of freedom quantized.

An alternative approach, originally developed by Dirac, is to quantize the entire space of degrees of freedom of classical theory, and only then to impose the constraints [100, 101, 102]. In Dirac quantization, states are initially determined from the full classical phase space; in quantum gravity, for instance, they are functionals $\Psi[g_{ij}]$ of the full spatial metric. The constraints act as operators on this auxiliary Hilbert space, and the physical Hilbert space consists of those states that are annihilated by the constraints, acted on by physical operators that commute with the constraints. For gravity, in particular, the Hamiltonian constraint (2.13) acting on states leads to a functional differential equation, the Wheeler–DeWitt equation [97, 285], whose solutions are the physical states.

In this chapter, we will consider several forms of Dirac quantization of (2+1)-dimensional general relativity, based on the first- and second-order formalisms. We shall see that the first-order version of the Wheeler–DeWitt equation is rather straightforwardly equivalent to the corresponding reduced phase space quantum theory, but that the second-order version, which involves a Hamiltonian constraint quadratic in the momenta, is considerably more problematic.

8.1 The first-order formalism

Let us first consider Dirac quantization of (2+1)-dimensional gravity in first-order form, with $\Lambda = 0$ for simplicity. As in previous chapters, we restrict our attention to spacetimes with topologies $[0, 1] \times \Sigma$, where Σ is a closed surface of genus $g \geq 1$. The canonical variables are then the spin connection $\omega_i{}^a$ and the frame $e_i{}^a$ on Σ, and the Poisson brackets (2.100) determine commutators

$$[e_i{}^a(x), \omega_{jb}(x')] = -\frac{i}{2}\epsilon_{ij}\delta_b^a\delta^2(x - x')$$

$$[e_{ia}(x), e_j{}^b(x')] = [\omega_{ia}(x), \omega_j{}^b(x')] = 0. \tag{8.1}$$

Choosing ω as our configuration space variable, we can represent e as a functional derivative

$$e_{ia} = -\frac{i}{2}\epsilon_{ij}\frac{\delta}{\delta\omega_j{}^a}. \tag{8.2}$$

Acting on wave functions $\Psi[\omega]$, the constraints (2.98) of chapter 2 then become the first-order Wheeler–DeWitt equation

$$\tilde{\mathscr{C}}^a\Psi[\omega] = \frac{1}{2}\epsilon^{ij}\left(\partial_i\omega_j{}^a - \partial_j\omega_i{}^a + \epsilon^{abc}\omega_{ib}\omega_{jc}\right)\Psi[\omega] = 0$$

$$\mathscr{C}^a\Psi[\omega] = \frac{i}{2}\left\{\partial_k\frac{\delta}{\delta\omega_{ka}} + \epsilon^{abc}\omega_{kb}\frac{\delta}{\delta\omega_k{}^c}\right\}\Psi[\omega] = 0, \tag{8.3}$$

and it is easy to check that the $ISO(2, 1)$ commutation relations (2.73), appropriately rescaled, are satisfied exactly.

The first of the constraints (8.3) involves no functional derivatives, and simply requires that $\Psi[\omega]$ have its support on flat $SO(2, 1)$ connections. The second constraint is then a statement of $SO(2, 1)$ gauge invariance. Indeed, the operator

$$1 + 2i\int_\Sigma d^2x\,\tau_a(x)\mathscr{C}^a(x)$$

generates the infinitesimal gauge transformation (2.66) of ω, so the condition $\mathscr{C}^a\Psi[\omega] = 0$ will be satisfied precisely when Ψ is invariant under 'small' gauge transformations.[*]

---

[*] Note that equation (8.3) does not require invariance under 'large' gauge transformations, transformations that cannot be smoothly deformed to the identity; these must be considered separately.

Physical states in Dirac quantization are thus gauge-invariant functionals of flat connections. This is exactly the description we obtained in chapter 6 from covariant canonical quantization. This simple equivalence occurs because the Chern–Simons formulation reduces (2+1)-dimensional gravity to a gauge theory, for which the constraints are relatively simple and well-behaved. In particular, it was crucial to our argument that the constraints were at most linear in the momenta $e_i{}^a$, leading to a first-order functional differential equation (8.3) that could be solved in closed form.

It is useful to make explicit a hidden assumption about the inner product on the space of physical wave functions. We began with a space of arbitrary functionals $\Psi[\omega]$ of the spin connection, upon which the natural inner product is the functional integral

$$\langle \Psi | \Phi \rangle = \int [d\omega] \Psi^*[\omega] \Phi[\omega]. \tag{8.4}$$

Our final physical states, on the other hand, differ in two important respects: they are gauge-invariant functionals, and they have their support only on flat connections. Both of these properties require changes in the inner product – the functional integral (8.4) must be gauge-fixed, and it must be appropriately restricted to flat connections.

The new inner product can be derived in three steps:

1. To gauge-fix the inner product, we can express an arbitrary connection ω in terms of a gauge-fixed connection $\bar{\omega}$ and an element $g \in SO(2,1)$: $\omega = {}^g\bar{\omega} = g^{-1}dg + g^{-1}\bar{\omega}g$. Then standard results from gauge theory tell us that, up to subtleties involving zero modes,

$$[d\omega] = [d\bar{\omega}][dg]\det D_{\bar{\omega}}, \tag{8.5}$$

where $D_{\bar{\omega}}$ is the gauge-covariant derivative with respect to the connection $\bar{\omega}$. As usual, we factor out the (infinite, field-independent) integral over gauge parameters g; the determinant $\det D_{\bar{\omega}}$ is then the usual Faddeev–Popov determinant.

2. To account for the fact that wave functions are restricted to flat connections, let us write

$$\Psi[\omega] = \psi[\bar{\omega}]\delta[\tilde{\mathscr{C}}^a[\bar{\omega}]], \qquad \Phi[\omega] = \phi[\bar{\omega}]\delta[\tilde{\mathscr{C}}^a[\bar{\omega}]],$$

$$\tag{8.6}$$

and define a new inner product

$$\langle\langle \Psi | \Phi \rangle\rangle = \int [d\bar{\omega}]\det D_{\bar{\omega}}\delta[\tilde{\mathscr{C}}^a[\bar{\omega}]]\psi^*[\bar{\omega}]\phi[\bar{\omega}], \tag{8.7}$$

where one of the delta functionals $\delta[\tilde{\mathscr{C}}^a[\omega]]$ has been omitted in order to avoid an overall infinite factor.

3. Finally, we can eliminate the delta functional in (8.7). Let $\bar{\omega}_0$ be a flat connection, and suppose that $\bar{\omega} = \bar{\omega}_0 + \epsilon$. It follows from the expression (8.3) for the constraints that to lowest order in ϵ,

$$\det D_{\bar{\omega}}\delta[\tilde{\mathscr{C}}^a[\omega]] = \delta[\epsilon], \tag{8.8}$$

where I have used the infinite-dimensional generalization of the identity

$$\int d^n x\, \delta^n[F(x)] = \left|\det\left(\frac{\partial F^i}{\partial x^j}\right)\right|^{-1}_{F=0}. \tag{8.9}$$

Changing integration variables in (8.7), we thus obtain

$$\langle\langle\Psi|\Phi\rangle\rangle = \int [d\bar{\omega}_0]\psi^*[\bar{\omega}_0]\phi[\bar{\omega}_0], \tag{8.10}$$

where the integral is now restricted to gauge-fixed, flat connections.[†]

In the end, we have obtained the naive inner product that we might have guessed from the start, with no troublesom Faddeev–Popov determinants. We shall see below that the issue is considerably more subtle in the metric formalism.

8.2 A quantum Legendre transformation

To quantize a classical theory, we must split the phase space into variables that we treat as 'positions' and conjugate variables that we treat as 'momenta'. Such a splitting is called a polarization, and it is rarely unique. We have chosen a polarization in which the connections ω are 'positions' and the triads e are 'momenta'. To obtain a picture closer to the second-order formalism, we might instead choose a 'frame representation', treating triads as configuration space variables and thus obtaining states that are functionals of the metric.

In such a polarization, the constraints are no longer first order in functional derivatives, and the resulting equations are considerably more complicated. This should not be surprising, since the connection representation reflects the structure of the phase space more closely than the frame representation. Indeed, as we saw in chapter 4, the classical phase space is a cotangent bundle with a base space parametrized by ω and

---

[†] Note that I have used two slightly different methods for gauge-fixing the two symmetries of the problem. The first, standard gauge-fixing process is essentially that described by Woodard in the usual approach to Dirac quantization [293]. The second is closely related to that described by Marolf in his 'refined algebraic quantization scheme' [186]; see reference [187] for an application to Euclidean quantum gravity.

fibers parametrized by e, making it far more natural to treat the ω as positions and the e as momenta.

On the other hand, a change of polarization can also be represented, at least formally, as a functional Fourier transform (or a 'quantum Legendre transformation'),

$$\tilde{\Psi}[e] = \int [d\omega] \exp\left\{2i \int_{\Sigma} e^a \wedge \omega_a\right\} \Psi[\omega], \qquad (8.11)$$

analogous to the transformation from position to momentum space in ordinary quantum mechanics. It is easy to see that such a transformation reproduces the representation (8.2) of the canonical operators:

$$e_{ia}\tilde{\Psi}[e] = \int [d\omega] \left(-\frac{i}{2}\epsilon_{ij}\frac{\delta}{\delta\omega_j{}^a} \exp\left\{2i \int_{\Sigma} e^a \wedge \omega_a\right\}\right) \Psi[\omega]$$

$$= \int [d\omega] \exp\left\{2i \int_{\Sigma} e^a \wedge \omega_a\right\} \frac{i}{2}\epsilon_{ij}\frac{\delta}{\delta\omega_j{}^a}\Psi[\omega], \qquad (8.12)$$

where the last equality comes from functional integration by parts. Such functional Fourier transforms have been studied in the context of conformal field theory, and a fair amount is known about their properties [264]. In particular, to make sense of a functional integral such as (8.11), one must fix the gauge, and the resulting Fadeev–Popov determinants can affect the definition of the wave function and the inner product. For the special case of Euclidean (2+1)-dimensional gravity with a positive cosmological constant, Ishibashi has investigated the functional integral (8.11), and has shown that the only effect of gauge-fixing is to finitely renormalize Newton's constant [163]. A related analysis has been carried out for (2+1)-dimensional supergravity [190]. Although this work is still preliminary, it suggests that it should be possible to define diffeomorphism-invariant 'ADM-type' wave functions $\tilde{\Psi}[e]$ that satisfy the appropriate Wheeler–DeWitt equation.

8.3 The second-order formalism

We now turn our attention to the metric formulation of (2+1)-dimensional gravity. The aim of this section is to apply Dirac quantization to the ADM decomposition developed in chapter 2, thus obtaining the standard second-order Wheeler–DeWitt equation. We shall see that the careful treatment of the momentum constraints requires the introduction of nonlocal terms that have not been appreciated in previous minisuperspace models, making Wheeler–DeWitt quantization rather problematic [59].[‡]

[‡] The existence of these nonlocal terms was first pointed out to me by Atsushi Higuchi and Bob Wald.

Much of the preliminary work we require has already been done in chapter 2. We begin with the decompositions (2.21) and (2.23) of the spatial metric and its canonically conjugate momentum, and with the constraints (2.27) and (2.28), which we now wish to represent as operators. To obtain the Wheeler–DeWitt equation, we must solve the momentum constraints (2.27) for Y_i, the momentum conjugate to the spatial diffeomorphisms; substitute the results into the Hamiltonian constraint (2.28); and then represent the remaining momenta as functional derivatives. The form of these functional derivatives is determined by the symplectic structure of equation (2.26):

$$\pi = -\frac{i}{2}\frac{\delta}{\delta\lambda}, \qquad p^{ij} = -i\frac{\delta}{\delta\tilde{g}_{ij}}. \tag{8.13}$$

The resulting operator version of the Hamiltonian constraint (2.28) is messy, but as usual, it simplifies when the spatial topology is that of a torus. As in chapter 3, let us choose a standard set of flat metrics $\tilde{g}_{ij}$ by setting

$$d\tilde{s}^2 = \tau_2^{-1}|dx + \tau dy|^2, \tag{8.14}$$

and let $\tilde{\nabla}$ denote the covariant derivative for the connection compatible with $\tilde{g}_{ij}$. If we assume for simplicity that the cosmological constant vanishes, a straightforward computation yields the equation

$$\widehat{\mathcal{H}}\Psi[\lambda,\tau] = 0 = \left\{ \frac{1}{8}\frac{\delta}{\delta\lambda}e^{-2\lambda}\frac{\delta}{\delta\lambda} + \frac{1}{2}e^{-2\lambda}\Delta_0 + 2\tilde{\Delta}\lambda - 2e^{-2\lambda}Y^i[\pi]\tilde{\Delta}Y_i[\pi] \right.$$
$$\left. + 2e^{-2\lambda}\tilde{\nabla}_i\left[\left(2p^{ij} + \tilde{\nabla}^i Y^j[\pi] + \tilde{\nabla}^j Y^i[\pi] - \tilde{g}^{ij}\tilde{\nabla}_k Y^k[\pi]\right) Y_j[\pi] \right] \right\}\Psi[\lambda,\tau], \tag{8.15}$$

where Δ_0 is the Laplacian of chapter 5 on the torus moduli space,

$$\Delta_0 = -\tau_2^2\left(\frac{\partial^2}{\partial\tau_1^2} + \frac{\partial^2}{\partial\tau_2^2} \right), \tag{8.16}$$

and $Y_i[\pi]$ is determined by the solution of the momentum constraints (2.27),

$$Y_i[\pi] = \frac{i}{4}\tilde{\Delta}^{-1}\left[e^{2\lambda}\tilde{\nabla}_i\left(e^{-2\lambda}\frac{\delta}{\delta\lambda} \right) \right]. \tag{8.17}$$

The Wheeler–DeWitt equation (8.15) should be compared to the corresponding reduced phase space Schrödinger equation (5.5)–(5.6). Two important differences are immediately apparent, both arising from the fact that the 'pure gauge' degrees of freedom π and Y^i are no longer fixed classically. The first is that λ is now a function, not a constant, so the $\tilde{\Delta}\lambda$ term in (2.28) no longer drops out. The second is that a complicated nonlocal term depending on $Y_i[\pi]$ is now present. This nonlocal term makes it extremely hard to find exact solutions of equation (8.15). As Henneaux first pointed out [148], such a nonlocal term already exists when the spatial topology is that of a two-sphere. We saw in chapter 3 that the classical field equations have at most a single solution for this topology, and the Hilbert space thus consists of one state, which one might hope to find in the Wheeler–DeWitt approach. Even in this simple case, however, the explicit solution of the Wheeler–DeWitt equation is not known.

Note also that the expression (8.15) is defined only up to operator ordering ambiguities, which are poorly understood. Ideally, one would like to find an ordering for which the commutator $[\widehat{\mathscr{H}}(x), \widehat{\mathscr{H}}(x')]$ is proportional to $\widehat{\mathscr{H}}$. At the time of this writing, no such ordering is known, again because of complications arising from nonlocal terms.

Now, the reduced phase space wave function of chapter 5 was a function of the mean curvature $Tr K$, not the scale factor λ. To make the comparison to (8.15) clearer, let us perform a functional Fourier transformation of $\Psi[\lambda, \tau]$:

$$\Psi[\lambda, \tau] = \int [dK] \exp\left\{-i \int d^2x\, K e^{2\lambda}\right\} \tilde{\Psi}[K, \tau]. \tag{8.18}$$

The Wheeler–DeWitt equation then becomes

$$\left\{\left(K\frac{\delta}{\delta K}\right)^2 + \Delta_0 + 2i\frac{\delta}{\delta K}\tilde{\Delta} \ln \frac{\delta}{\delta K} - 4Y^i[K]\tilde{\Delta}Y_i[K] \right. \tag{8.19}$$

$$\left. + 4\tilde{\nabla}_i\left[\left(2p^{ij} + \tilde{\nabla}^i Y^j[K] + \tilde{\nabla}^j Y^i[K] - \tilde{g}^{ij}\tilde{\nabla}_k Y^k[K]\right) Y_j[K]\right]\right\}\tilde{\Psi}[K, \tau] = 0$$

where now

$$Y_i[K] = \frac{i}{2}\tilde{\Delta}^{-1}\left[\tilde{\nabla}_i K \frac{\delta}{\delta K}\right]. \tag{8.20}$$

If we could set K to a constant, $Y_i[K]$ would vanish, and this expression would be close to the square of the Schrödinger equation (5.5)–(5.6). We cannot do so, however, and it is not at all clear that the two quantizations can be made equivalent. Suppose, for example, that we look for solutions of the Wheeler–DeWitt equation that depend only on constant values of K; that is, let us set

$$\Psi_{trial}[\lambda, \tau] = \int dT \exp\left\{-iT \int d^2x\, e^{2\lambda}\right\} \tilde{\psi}_0(T, \tau). \tag{8.21}$$

From equation (8.17), we have

$$\tilde{\Delta} Y_i \, \Psi_{trial}[\lambda, \tau] = 0, \tag{8.22}$$

so the nonlocal terms in the Wheeler–DeWitt equation drop out. Unfortunately, however, the remaining terms require that

$$\int dT \exp\left\{-iT \int d^2x \, e^{2\lambda}\right\} \left[-\frac{1}{2} T^2 e^{2\lambda} + \frac{1}{2} e^{-2\lambda} \Delta_0 + 2\tilde{\Delta}\lambda \right] \tilde{\psi}_0(T, \tau) = 0, \tag{8.23}$$

which cannot be satisfied by a λ-independent function $\tilde{\psi}_0[T, \tau]$.

With a bit of thought, the problem with the ansatz (8.21) becomes clear. The reduced phase space wave functions of chapter 5 were defined on a particular set of time slices $Tr K = -T$. A Wheeler–DeWitt wave functional, on the other hand, is a functional of Wheeler's 'many-fingered time' – it does not depend on a particular choice of time coordinate, but determines the wave function on an arbitrary slice. An ansatz that restricts the wave functional to a particular slicing cannot hope to capture the full content of the Wheeler–DeWitt equation.

This analysis suggests another strategy, however: we can start with an exact solution of the full Wheeler–DeWitt equation (8.15), restrict it to a slice $Tr K = -T$, and hope that the restricted wave function might be related to those of chapter 5. That is, suppose $\Psi[\lambda, \tau]$ is a solution of equation (8.15) (whose functional form we do not yet know), and let

$$\hat{\psi}_0(T, \tau) = \int [d\lambda] \exp\left\{iT \int d^2x \, e^{2\lambda}\right\} \Psi[\lambda, \tau]. \tag{8.24}$$

We can obtain information about $\hat{\psi}_0$ by inserting the Wheeler–DeWitt equation, in the form $e^{2\lambda} \widehat{\mathscr{H}} \Psi = 0$, into (8.24) and functionally integrating by parts to move the functional derivatives $\delta/\delta\lambda$ to the exponent. It is easy to see that the troublesome nonlocal terms are again eliminated: for any ζ^i,

$$\int [d\lambda] \exp\left\{iT \int d^2x \, e^{2\lambda}\right\} \zeta^i Y_i[\pi] \Psi[\lambda, \tau]$$

$$= -\int [d\lambda] \left(\zeta^i Y_i[\pi] \exp\left\{iT \int d^2x \, e^{2\lambda}\right\} \right) \Psi[\lambda, \tau] = 0 \tag{8.25}$$

by (8.17). A straightforward computation then shows that

$$\left[\left(T \frac{\partial}{\partial T} \right)^2 + \Delta_0 \right] \hat{\psi}_0(T, \tau) = \tag{8.26}$$

$$\int [d\lambda] \exp\left\{iT \int d^2x \, e^{2\lambda}\right\} \left[T^2 \left\{ e^{4\lambda} - \left(\int d^2x \, e^{2\lambda} \right)^2 \right\} + 4e^{2\lambda} \tilde{\Delta}\lambda \right] \Psi[\lambda, \tau].$$

The left-hand side of this equation is to be compared to the reduced phase space Schrödinger equation of chapter 5,

$$iT\frac{\partial}{\partial T}\psi_{red} = \Delta_0^{1/2}\psi_{red}. \tag{8.27}$$

If the Wheeler–DeWitt wave function $\Psi[\lambda,\tau]$ had its support on spatially constant scale factors λ, the right-hand side of (8.26) would vanish, and we would obtain the square of this Schrödinger equation. Once again, however, there seems to be no natural way to restrict $\Psi[\lambda,\tau]$ to such constant λ, and the Wheeler–DeWitt equation appears to be inequivalent to reduced phase space quantization.

As we saw in the first section of this chapter, however, there is more to the story: we have not yet defined the inner product on the space of solutions of the Wheeler–DeWitt equation. The need for gauge-fixing is less obvious now than it was in the first-order formalism, but as Woodard has argued [293], the inner product

$$\langle\Psi|\Phi\rangle = \int [d\lambda]\int \frac{d^2\tau}{\tau_2{}^2}\Psi^*[\lambda,\tau]\Phi[\lambda,\tau] \tag{8.28}$$

will again diverge for solutions of the Wheeler–DeWitt equation unless the transformations generated by the Hamiltonian constraint are gauge-fixed.

In particular, we can gauge-fix the inner product to the York gauge,

$$\chi = \pi - Te^{2\lambda} = 0, \tag{8.29}$$

in effect imposing the $TrK = -T$ time-slicing on the inner product rather than on the classical phase space. The Faddeev–Popov determinant for this gauge is proportional to $\det[\widehat{\mathscr{H}},\chi]$, and may be computed to be

$$v^2[\lambda,\tau] = \det\left|\tilde{\Delta} - \frac{1}{2}T^2e^{2\lambda} - \frac{1}{2}e^{-2\lambda}\Delta_0\right|_{\chi=0}, \tag{8.30}$$

up to operator ordering ambiguities. Note that unlike the more familiar Faddeev–Popov determinants of gauge theory, $v[\lambda,\tau]$ is an operator, depending explicitly on the Laplacian Δ_0 on moduli space.

The gauge-fixed form of the inner product (8.28) is now easy to find: up to ordering ambiguities, it is

$$\langle\langle\Psi|\Phi\rangle\rangle = \int \frac{d^2\tau}{\tau_2{}^2}\hat{\psi}^*(T,\tau)\hat{\phi}(T,\tau), \tag{8.31}$$

where

$$\hat{\psi}(T,\tau) = \int [d\lambda]v[\lambda,\tau]\exp\left\{iT\int d^2x\, e^{2\lambda}\right\}\Psi[\lambda,\tau]. \tag{8.32}$$

This effective wave function is almost the wave function $\hat{\psi}_0$ of (8.24), but it differs because of the presence of the Faddeev–Popov determinant. In reference [59], it was shown that the effect of this determinant is to cause the wave function to be peaked near the solutions of the classical constraints. Very little more is known, however. In particular, the question of whether the wave function (8.32) actually satisfies the reduced phase space Schrödinger equation (8.27) remains unanswered, and the equivalence of reduced phase space and Dirac quantization remains unproven.

8.4 Perturbation theory

In view of the difficulty in finding exact solutions to the Wheeler–DeWitt equation, it is natural to look for perturbative methods. One possible starting point is an expansion in powers of Newton's constant G. This expansion is most easily described for the topology $[0, 1] \times S^2$; although this topology gives a trivial reduced phase space quantum theory, the Wheeler–DeWitt equation is already complicated enough to make the example instructive.

The Wheeler–DeWitt equation analogous to (8.15) for $[0, 1] \times S^2$, with a cosmological constant included and factors of G restored, is

$$\widehat{\mathscr{H}}\Psi[\lambda] = \left\{ \frac{(16\pi G)^2}{8} \frac{1}{\sqrt{\tilde{g}}} \frac{\delta}{\delta\lambda} e^{-2\lambda} \frac{\delta}{\delta\lambda} + 2\sqrt{\tilde{g}} \left[\tilde{\Delta}\lambda - \frac{1}{2} \right] \right.$$
$$\left. + 2\sqrt{\tilde{g}}\Lambda e^{2\lambda} + \textit{nonlocal terms} \right\} \Psi[\lambda, \tau] = 0. \quad (8.33)$$

Let us write a trial wave function

$$\Psi_0[\lambda] = \exp\left\{ \frac{i\alpha_1}{(16\pi G)^2} \int d^2x \sqrt{\tilde{g}} e^{2\lambda} \right\}. \quad (8.34)$$

This functional satisfies

$$e^{-2\lambda} \frac{\delta\Psi_0}{\delta\lambda} = \frac{2i\alpha_1}{(16\pi G)^2} \sqrt{\tilde{g}}\Psi_0[\lambda], \quad (8.35)$$

which implies from (8.17) that $Y_i\Psi_0 = 0$, so to this order the nonlocal terms in (8.33) drop out. It is then easily checked that

$$\widehat{\mathscr{H}}\Psi_0[\lambda] = \left\{ \left(2\Lambda - \frac{\alpha_1{}^2}{2(16\pi G)^2} \right) \sqrt{\tilde{g}}e^{2\lambda} + 2\sqrt{\tilde{g}} \left(\tilde{\Delta}\lambda - \frac{1}{2} \right) \right\} \Psi_0[\lambda]. \quad (8.36)$$

If we take the cosmological constant to be of order $1/G^2$, the Wheeler–DeWitt equation will then be satisfied to lowest order if

$$\alpha_1 = \pm 32\pi G\Lambda^{1/2} = \pm\frac{2}{\ell}, \quad (8.37)$$

where ℓ is the dimensionless radius of curvature introduced at the end of chapter 1.

At next order, Ψ must include a term that cancels the curvature term $\tilde{\Delta}\lambda - 1/2$ in (8.36). It is not hard to check that a wave functional

$$\Psi_1[\lambda] = \exp\left\{\frac{i\alpha_1}{(16\pi G)^2}\int d^2x\sqrt{\tilde{g}}e^{2\lambda} + i\alpha_2\int d^2x\sqrt{\tilde{g}}\left[\lambda\tilde{\Delta}\lambda - \lambda\right]\right\} \tag{8.38}$$

will do the trick, with

$$\alpha_2 = sgn(\alpha_1)\ell. \tag{8.39}$$

The nonlocal terms cannot be neglected at this order, but they produce terms proportional to $\delta^2(0)$ that arguably can be removed by an appropriate operator ordering.

In terms of the full spatial curvature, the second term in (8.38) is of the form

$$\int d^2x\sqrt{g(x)}\int d^2x'\sqrt{g(x')}R(x)\Delta^{-1}(x,x')R(x').$$

Using somewhat different methods, Banks, Fischler, and Susskind have computed the next term, which involves three factors of the curvature and two Green's functions [24]. They argue that the effective expansion parameter is $(\Lambda A)^{-1}$, where A is the area of the spatial surface upon which the wave functional is being evaluated; the expansion should therefore be good for large universes.

Unfortunately, approximation techniques of this type become quite a bit more complicated for spacetimes with nontrivial spatial topology, and they have yet to lead to results that can be easily compared to other forms of quantization. Consider, for example, the torus universe, now with a cosmological constant Λ. The first-order approximation (8.34) remains good. At the next order, however, the Laplacian Δ_0 in equation (8.15) can no longer be neglected, and a term of the form

$$\alpha_2\Delta_0(\lambda\tilde{\Delta}\lambda) = -2\alpha_2\lambda\tilde{\Delta}\lambda \tag{8.40}$$

appears in the Wheeler–DeWitt equation. There is no obvious way to cancel (8.40) without making the approximation considerably more complicated. The nonlocal terms in the Wheeler–DeWitt equation – the terms in (8.15) involving $Y_i[\pi]$ – are also poorly understood, as is the related problem of operator ordering.

The Wheeler–DeWitt equation is the starting point for many of the most widely used approaches to quantum gravity. The results of this chapter should be read as a warning: the full Wheeler–DeWitt equation is considerably more complicated than is generally appreciated. In particular, minisuperspaces are misleading models, which miss the nonlocal terms that have given us so much trouble and which evade the problems of gauge-fixing the inner product that must be confronted in the full theory.

9
Lorentzian path integrals

The last four chapters have all dealt with approaches to quantum gravity that fall under the broad heading of canonical quantization. An alternative approach starts with the Feynman path integral, the 'sum over histories'. Like canonical quantization, path integration does not always lead to a single, unique quantum theory. In this chapter I will describe several path integral techniques in (2+1)-dimensional gravity and compare the results to those of canonical quantization.

In an important sense, path integral methods are less precise than those of canonical quantization. The infinite-dimensional 'integral' over histories can rarely be rigorously defined; typically, we do not even know what class of paths to include. Path integrals may seem to avoid operator-ordering problems, since the integration variables are commuting classical fields rather than operators, but ordering ambiguities reappear in the guise of ambiguities in the integration measure. In quantum gravity, however, path integral formulations also have an important advantage: we need no longer assume that spacetime has the topology $[0, 1] \times \Sigma$, and can compute topology-changing amplitudes.

The focus of this chapter is the Lorentzian path integral, the sum over spacetime histories with metrics of signature $(- + +)$. The analytically continued 'Euclidean' path integral – the sum over positive definite metrics – has also seen extensive use in (3+1)-dimensional quantum cosmology; I will describe its application in 2+1 dimensions in the next chapter.

9.1 Path integrals and ADM quantization

The Feynman path integral is formally defined as either

$$\langle q_f, t_f \mid q_i, t_i \rangle = \int [dq] e^{iI[q]} \tag{9.1}$$

143

or

$$\langle q_f, t_f \mid q_i, t_i \rangle = \int [dq][dp] e^{iI[q,p]} \qquad (9.2)$$

where the (formal) integration is over all histories in configuration space (equation (9.1)) or phase space (equation (9.2)) that interpolate between specified initial and final data q_i and q_f. The path integral gives a transition amplitude between the states $|q_i\rangle$ and $|q_f\rangle$; for simple enough systems, the result is equivalent to the amplitude coming from canonical quantization. To apply this formalism to gravity, we must specify geometric data on an initial spatial slice Σ_i and a final slice Σ_f, and sum over intermediate geometries on spacetimes M with boundary $\partial M = \Sigma_i \cup \Sigma_f$. If Σ_i and Σ_f are diffeomorphic, one may either restrict M to have the topology $[0,1] \times \Sigma$ or sum over all interpolating topologies. If Σ_i and Σ_f are topologically distinct, on the other hand, there seems to be no reasonable way to restrict the intermediate topologies.

It should come as no surprise that the role of time in the gravitational path integral is rather obscure. In ordinary quantum mechanics, the initial and final times appear as part of the specification of the histories to be summed over. In quantum gravity, this is no longer possible, since there is no external time variable. Instead, information about time must be included in the specification of the boundary data itself. Different time-slicings then correspond in part to different choices of boundary data. The York time-slicing of chapter 2, for instance, is obtained if we specify the mean extrinsic curvature TrK and the conformal part of the spatial metric on Σ_i and Σ_f and choose a suitable gauge-fixing for the interpolating spacetime. As we shall see later in this chapter, covariant canonical quantization corresponds to the specification of the holonomies of the spin connection ω on Σ_i and Σ_f.

There is a fairly easy heuristic argument that the path integral for a (2+1)-dimensional spacetime with the topology $[0,1] \times \Sigma$ should reproduce the results of canonical quantization. In the ADM action (2.12), the lapse and shift functions N and N^i appear as Lagrange multipliers, and their integration leads to delta functionals that impose the constraints.* We are then left with a path integral for the reduced phase space action (2.36) over a relatively simple space of physical degrees of freedom. For such a finite-dimensional system, the equivalence of path integration and canonical quantization is more or less standard. In the first-order formalism, the result is even simpler: integrals over $e_t{}^a$ and $\omega_t{}^a$ in the canonical action (2.97) again give delta functionals that impose the constraints, and the remaining finite-dimensional path integral involves an action with a

---

* There are additional subtleties involving the range of integration over the lapse function; I refer the reader to reference [254] for a detailed discussion.

vanishing Hamiltonian. Of course, this argument is not yet rigorous: when we change variables to those appropriate for the reduced phase space, we will introduce a Jacobian, the Faddeev–Popov determinant, which must be treated carefully.

Let us examine this argument in more detail, starting with the phase space path integral in the ADM formalism [61]. The canonical action (2.12) is

$$I_{grav} = \int d^3x \sqrt{-^{(3)}g} \, ^{(3)}R = \int dt \int_\Sigma d^2x (\pi^{ij}\dot{g}_{ij} - N^i\mathcal{H}_i - N\mathcal{H}),$$
(9.3)

and the path integral (9.2) is

$$Z = \int [d\pi^{ij}][dg_{ij}][dN^i][dN] \exp\{iI_{grav}[\pi, g]\}.$$
(9.4)

We know from chapter 2 that the action has a gauge symmetry, essentially the group of diffeomorphisms, which is generated by the constraints $\mathcal{H}_\mu = (\mathcal{H}, \mathcal{H}_i)$ and which must be gauge-fixed. The general theory of path integrals with gauge symmetries tells us that for gauge conditions $\chi^\mu = 0$, the path integral should become

$$Z = \int [d\pi^{ij}][dg_{ij}][dN^i][dN]\delta[\chi^\mu] \det|\{\mathcal{H}_\mu, \chi^\nu\}| \exp\{iI_{grav}[\pi, g]\},$$
(9.5)

where $\{ , \}$ denotes the Poisson bracket [149].

To simplify the computation, it is again useful to parametrize the spatial metric and its conjugate momentum, using a decomposition similar to that of equations (2.21)–(2.23). We choose a family $\tilde{g}_{ij}(m_\alpha)$ of constant curvature metrics, parametrized by moduli m_α, with

$$g_{ij} = e^{2\lambda}f^*\tilde{g}_{ij},$$
(9.6)

and set

$$\delta g_{ij} = 2(\delta\lambda)g_{ij} + (\hat{P}(\delta\xi))_{ij} + \delta m_\alpha T^{\alpha\beta}\Psi_{(\beta)ij},$$

$$\pi^{ij} = \frac{1}{2}g^{ij}\pi + \sqrt{g}(\hat{P}Y)^{ij} + \sqrt{g}\,\hat{p}^\alpha\Psi_{(\alpha)}{}^{ij}.$$
(9.7)

Here $\hat{P}$ is the operator

$$(\hat{P}\xi)_{ij} = \nabla_i\xi_j + \nabla_j\xi_i - g_{ij}\nabla_k\xi^k.$$
(9.8)

(Note that in contrast to the definition (2.22), the derivative in (9.8) is now ∇_i, the covariant derivative compatible with the full spatial metric

g_{ij}. This change will simplify future notation.) The $\Psi_{(\alpha)}$ are a basis for $ker\,\hat{P}^\dagger$, that is,

$$\nabla_j\Psi_{(\alpha)}{}^{ij} = 0, \quad g_{ij}\Psi_{(\alpha)}{}^{ij} = 0. \tag{9.9}$$

They are thus a basis for the space of transverse traceless tensors, or quadratic differentials. The operator $T^{\alpha\beta}$ is new: it is the orthogonal projection of a modular deformation of g_{ij} onto $ker\,\hat{P}^\dagger$,

$$T^{\alpha\beta} = \langle\chi^{(\alpha)}, \Psi_{(\gamma)}\rangle\langle\Psi_{(\gamma)}, \Psi_{(\beta)}\rangle^{-1}, \quad \text{with} \quad \chi_{ij}^{(\alpha)} = \frac{\partial g_{ij}}{\partial m_\alpha}, \tag{9.10}$$

where the inner product in this expression is defined below in equation (9.12). This projection is introduced because it is important in a path integral computation to keep track of orthogonality: if we wish to avoid unnecessary Jacobians, we must make sure that the decomposition of integration variables is an orthogonal one. It is easy to check that the constraints become

$$\mathcal{H}_i = \sqrt{g}(\hat{P}^\dagger\hat{P}Y)_i - \nabla_i\pi$$
$$\mathcal{H} = -\frac{1}{2}\frac{1}{\sqrt{\tilde{g}}}e^{-2\lambda}\pi^2 + 2\sqrt{\tilde{g}}\left(\tilde{\Delta}\lambda - \frac{k}{2}\right)$$
$$+ \sqrt{\tilde{g}}\tilde{g}^{ik}\tilde{g}^{jl}e^{-2\lambda}\left((\hat{P}Y)_{ij} + \hat{p}^\alpha\Psi_{(\alpha)ij}\right)\left((\hat{P}Y)_{kl} + \hat{p}^\beta\Psi_{(\beta)kl}\right), \tag{9.11}$$

where I have taken $\Lambda = 0$ for simplicity.

Our next step is to transform the integration variables in the path integral (9.5) from (g_{ij}, π^{ij}) to $(m_\alpha, \lambda, \xi, \hat{p}^\alpha, \pi, Y^i)$. To find the Jacobian J for this transformation, we can use an approach borrowed from string theory [203]. We define the inner products

$$\langle\delta g, \delta g\rangle = \int_\Sigma d^2x\,\sqrt{g}g^{ij}g^{kl}\delta g_{ik}\delta g_{jl}$$
$$\langle\delta\xi, \delta\xi\rangle = \int_\Sigma d^2x\,\sqrt{g}g^{ij}\delta\xi_i\delta\xi_j$$
$$\langle\delta\lambda, \delta\lambda\rangle = \int_\Sigma d^2x\,\sqrt{g}\,(\delta\lambda)^2 \tag{9.12}$$

on the tangent space to the space of metrics on Σ, and consider the simple Gaussian path integral

$$1 = \int [d(\delta g_{ij})]\,e^{i\langle\delta g,\delta g\rangle} \tag{9.13}$$
$$= \int [d(\delta m)]\,[d(\delta\lambda)][d(\delta\xi)]J_g\,e^{i\langle\delta\xi,\hat{P}^\dagger\hat{P}\delta\xi\rangle}e^{8i\langle\delta\lambda,\delta\lambda\rangle}e^{i\delta m_\alpha\delta m_\beta\,T^{\alpha\gamma}T^{\beta\delta}\langle\Psi_{(\gamma)},\Psi_{(\delta)}\rangle}.$$

As discussed above, the orthogonal projectors $T^{\alpha\beta}$ ensure that there are no cross-terms in the last line of this equation. Evaluating the integrals over δm_α, $\delta\lambda$, and $\delta\xi$, we find that the Jacobian J_g is

$$J_g = \det|\hat{P}^\dagger \hat{P}|^{1/2} \det|T| \det|\langle \Psi_{(\alpha)}, \Psi_{(\beta)} \rangle|^{1/2}. \qquad (9.14)$$

The π^{ij} integral gives a similar Jacobian,

$$J_\pi = \det|\hat{P}^\dagger \hat{P}|^{1/2} \det|\langle \Psi_{(\alpha)}, \Psi_{(\beta)} \rangle|^{1/2}, \qquad (9.15)$$

which combines with J_g to give a total Jacobian – that is, a total Faddeev–Popov determinant –

$$J = \det|\hat{P}^\dagger \hat{P}| \det|\langle \chi^{(\alpha)}, \Psi_{(\beta)} \rangle|. \qquad (9.16)$$

Strictly speaking, the Jacobian J was defined on the tangent space to phase space – that is, we integrated linear variations δg and $\delta\pi$ around a fixed point (g, π) in phase space. But viewed as a function on phase space, J is also the correct Jacobian for the change of variables in the path integral (9.5).

If we now change variables and integrate over N^μ, the path integral becomes

$$Z = \int [d^n \hat{p}] \, [d^n m] \det|\langle \chi^{(\alpha)}, \Psi_{(\beta)} \rangle| \qquad (9.17)$$

$$\int [d(\pi/\sqrt{g})][d\lambda][dY][d\xi] \det|\hat{P}^\dagger \hat{P}| \, \delta[\chi^\mu] \, \delta[\mathcal{H}_\nu/\sqrt{g}] \det|\{\mathcal{H}_\mu, \chi^\nu\}| e^{iI_{\text{grav}}}.$$

The integral over Y_i can now be evaluated, using the delta functional $\delta[\mathcal{H}_i/\sqrt{g}]$ and the identity (8.9). From (9.11), we see that

$$\int [dY_i] \det|\hat{P}^\dagger \hat{P}| \, \delta[\mathcal{H}_i/\sqrt{g}] = 1, \qquad (9.18)$$

where from now on, Y_i will mean the value determined by the solution of the constraints $\mathcal{H}_i = 0$.

To proceed further, we choose the York time gauge of chapter 2; that is, we choose a gauge-fixing functional

$$\chi^0 = \pi/\sqrt{g} - T = 0. \qquad (9.19)$$

Using the symplectic structure (2.26) to evaluate the Poisson brackets in equation (9.17), it is not hard to check that $\det|\{\mathcal{H}_\mu, \chi^\nu\}|$ factors into $\det|\{\mathcal{H}_i, \chi^j\}| \det|\{\mathcal{H}_0, \chi^0\}|$, and that

$$\det|\{\mathcal{H}_0, \chi^0\}|$$

$$= \det\left| e^{-2\lambda} \left(-\tilde{\Delta} + \frac{T^2}{2} e^{2\lambda} + e^{-2\lambda} \hat{p}^\alpha \hat{p}^\beta \tilde{g}^{ij} \tilde{g}^{kl} \Psi_{(\alpha)ik} \Psi_{(\beta)jl} \right) \right|. \qquad (9.20)$$

The only remaining ξ-dependent terms in the path integral are

$$\int [d\xi]\delta[\chi^i]\det|\{\mathcal{H}_i,\chi^j\}| = 1, \tag{9.21}$$

and (9.17) simplifies to

$$Z = \int [d^n\hat{p}]\,[d^n m]\det|\langle\chi^{(\alpha)},\Psi_{(\beta)}\rangle|\int [d\lambda]\delta[\mathcal{H}/\sqrt{g}]\det|\{\mathcal{H}_0,\chi^0\}|e^{iI_{grav}}. \tag{9.22}$$

We can now use the remaining delta functional to evaluate the integral over λ. As we saw in chapter 2, the equation $\mathcal{H} = 0$ has a unique solution $\lambda = \bar{\lambda}(\hat{p}, m, T)$ in the York gauge, so

$$\delta[\mathcal{H}/\sqrt{g}] = \det\left|\frac{\delta}{\delta\lambda}(\mathcal{H}/\sqrt{g})\right|^{-1}\delta[\lambda - \bar{\lambda}]. \tag{9.23}$$

The functional derivative in this equation may be calculated from (9.11); it simply cancels the remaining determinant $\det|\{\mathcal{H}_0,\chi^0\}|$ in the path integral (9.22). We thus obtain

$$Z = \int [d^n\hat{p}]\,[d^n m]\det|\langle\chi^{(\alpha)},\Psi_{(\beta)}\rangle|\,e^{i\bar{I}_{grav}}, \tag{9.24}$$

where $\bar{I}$ is the action evaluated at the solution of the constraints.

Finally, let us consider the remaining determinant $\det|\langle\chi^{(\alpha)},\Psi_{(\beta)}\rangle|$. In the chapter, we defined the reduced phase space momenta $\hat{p}_\alpha$ by the decomposition (9.7). In chapter 2, on the other hand, we used a slightly different parametrization,

$$p^\alpha = \int_\Sigma d^2x\, e^{2\lambda}(\pi^{ij} - \frac{1}{2}g^{ij}\pi)\frac{\partial\tilde{g}_{ij}}{\partial m_\alpha} = \langle\hat{p}^\beta\Psi_{(\beta)},\chi^{(\alpha)}\rangle = \langle\chi^{(\alpha)},\Psi_{(\beta)}\rangle\hat{p}^\beta. \tag{9.25}$$

Thus by changing variables to p^α, we can finally write the path integral (9.24) as

$$Z = \int [d^n p]\,[d^n m]\,e^{i\bar{I}_{grav}[p,m]}. \tag{9.26}$$

This is precisely the path integral we would obtain from the reduced phase space quantization of chapter 2. But (9.26) is now an ordinary quantum mechanical path integral, and standard arguments show that it gives amplitudes equivalent to those of the corresponding canonical quantum theory.

If the spatial slice Σ has the topology of a torus, several added complications appear, arising from the fact that the flat torus admits Killing vectors [239]. First, a Killing vector χ^i is a zero-mode of the operator $\hat{P}^\dagger\hat{P}$, so the determinant $\det|\hat{P}^\dagger\hat{P}|$ in (9.17) is identically zero. This problem may be traced back to the integral (9.13) used to derive the Jacobian (9.14). In the presence of $\hat{P}^\dagger\hat{P}$ zero-modes, we must split the integration over $\delta\xi$ into an integral of the form

$$\int [d(\delta\xi)] = \int' [d(\delta\xi)] \int_{\substack{\text{Killing} \\ \text{vectors}}} [d\chi], \qquad (9.27)$$

where the prime denotes integration over the space spanned by modes of $\hat{P}^\dagger\hat{P}$ with nonzero eigenvalues. A similar correction is necessary in the integral used to determine the Jacobian (9.15). Equation (9.16) thus becomes[†]

$$J = V_K^{-1} \det'|\hat{P}^\dagger\hat{P}| \det|\langle\chi^{(\alpha)}, \Psi^{(\beta)}\rangle|, \qquad (9.28)$$

where V_K is the volume of the subgroup of diffeomorphisms generated by Killing vectors. This volume may be understood heuristically by noting that the gauge-fixed path integral involves a factor of the reciprocal of the volume of the gauge group. In the absence of Killing vectors, this volume cancels the gauge redundancy in the original path integral. When Killing vectors are present, however, a factor of V_K remains in the denominator: a Killing vector χ does *not* give an extra, redundant contribution to the path integral, since g_{ij} and $g_{ij} + \mathscr{L}_\chi g_{ij}$ are no longer distinct configurations.

A careful analysis, however, reveals an additional factor of V_K. Recall that the delta functional $\delta[\mathscr{H}_i/\sqrt{g}]$ in equation (9.17) came from an integral over the shift vector N^i. If N^i is a Killing vector, however, it makes no contribution to this delta function: the relevant term in the action is

$$2\int dt \int_\Sigma d^2x N^i \mathscr{H}_i = -\int dt \int_\Sigma d^2x (\nabla^i N^j + \nabla^j N^i)\pi_{ij}, \qquad (9.29)$$

which vanishes for Killing vectors. To integrate over N^i, we must again split the integral as in (9.27). The result is a term of the form $V_K\delta[\mathscr{H}_i/\sqrt{g}]$, and the factor of V_K exactly cancels the corresponding term in the Jacobian (9.28). The upshot is that the reduced phase space path integral (9.26) remains correct, and the equivalence with canonical quantization is maintained.

9.2 Covariant metric path integrals

Like the reduced phase space quantum theory of chapter 5, the path integral of the preceding section is not manifestly covariant, but requires

---

[†] A primed determinant $\det'\mathcal{O}$ denotes a determinant restricted to the space spanned by modes of $\mathcal{O}$ with nonvanishing eigenvalues. Equivalently, $\det'\mathcal{O}$ is the product of the nonzero eigenvalues of $\mathcal{O}$, defined, for instance, by zeta function regularization.

an explicit choice of time-slicing and a corresponding decomposition of fields. This is reminiscent of Coulomb or axial gauge in quantum electrodynamics. It is natural to ask whether a covariant path integral, analogous to the Landau or Feynman gauge path integral in QED, can be formulated for gravity. This is an important problem: the ADM methods of the preceding section are valid only for spacetimes with the topology $[0, 1] \times \Sigma$, and cannot be used to discuss such issues as topology change.

We shall see in the next section that a covariant path integral treatment is possible in the first-order formalism. In the second-order formalism, less is known. However, the change of variables analogous to (9.7) has been worked out, and the resulting Jacobians have been computed [193].

Let M be a spacetime manifold, with a topology not necessarily of the form $[0, 1] \times \Sigma$, and let $g_{\mu\nu}$ denote a metric on M. Our first step is to find a parametrization of the metric analogous to that of equation (2.21) of chapter 2. For positive definite metrics – 'Riemannian' to mathematicians, 'Euclidean' in most of the physics literature – a decomposition very closely analogous to (2.21) can be found: the Yamabe conjecture, proved by Schoen in 1984, shows that any such metric is conformal to one of constant scalar curvature $R = 0$ or $R = \pm 1$ [236, 295]. Assuming the same is true for Lorentzian metrics,[‡] we can write

$$g_{\mu\nu} = e^{2\sigma} f^* \tilde{g}_{\mu\nu}, \tag{9.30}$$

where f is a spacetime diffeomorphism and $\tilde{g}_{\mu\nu}$ is a metric of constant scalar curvature. The gravitational action then becomes

$$I[\tilde{g}, \sigma] = \int_M d^3x \sqrt{-g}\, R = \int_M d^3x \sqrt{-\tilde{g}}\, e^{\sigma} \left[\tilde{R} - 2\tilde{g}^{\mu\nu} \partial_\mu \sigma \partial_\nu \sigma \right], \tag{9.31}$$

which for fixed $\tilde{g}_{\mu\nu}$ is the action of a scalar field in a curved background.

The infinitesimal decomposition analogous to (9.7) is now

$$\delta g_{\mu\nu} = 2(\delta\sigma)g_{\mu\nu} + (L(\delta\xi))_{\mu\nu} + \delta h_{\mu\nu},$$
$$g^{\mu\nu}\delta h_{\mu\nu} = 0, \qquad (L^\dagger \delta h)_\mu = 0, \tag{9.32}$$

where the operator L is the three-dimensional analog of (9.8),

$$(L\xi)_{\mu\nu} = \nabla_\mu \xi_\nu + \nabla_\nu \xi_\mu - \frac{2}{3} g_{\mu\nu} \nabla_\rho \xi^\rho. \tag{9.33}$$

Unlike a surface, a three-manifold M has an infinite-dimensional 'moduli space' – there are infinitely many transverse traceless deformations

---

[‡] Much of the mathematical manipulation that follows is, strictly speaking, valid only for Euclidean metrics. When necessary, I will assume that the path integral is defined by continuation to Euclidean signature.

$h_{\mu\nu}$. Nevertheless, much of the analysis of the preceding section remains unchanged. We can again define an inner product

$$\langle \delta g, \delta g \rangle_g = \int_M d^3x \sqrt{-g} g^{\mu\rho} g^{\nu\sigma} \delta g_{\mu\nu} \delta g_{\rho\sigma}, \qquad (9.34)$$

where the subscript g is meant to emphasize that indices are contracted with g and not $\tilde{g}$. As before, we can change integration variables from g to $(\tilde{g}, \sigma)$, determining the new integration measure by demanding that

$$\int [d(\delta g_{\mu\nu})] e^{i\langle \delta g, \delta g \rangle_g} = 1. \qquad (9.35)$$

The resulting Jacobian is identical to that of (9.14), and the measure becomes

$$[dg] = [d\sigma][d\xi][d\tilde{g}] \, \text{det}'|L^\dagger L|^{1/2} \frac{\text{det}|\langle \chi^{(\alpha)}, \Psi_{(\beta)} \rangle_g|}{\text{det}|\langle \Psi_{(\alpha)}, \Psi_{(\beta)} \rangle_g|^{1/2}}, \qquad (9.36)$$

where the $\Psi_{(\alpha)}$ are a basis for the kernel of $L^\dagger$ and the $\chi^{(\alpha)}$ are a basis for the tangent space to the Yamabe slice $R = \text{const}$.

In contrast to the canonical path integral, there are no additional changes of variable to cancel Faddeev–Popov determinants in (9.36), and they must be included in any computation. To take full advantage of the decomposition (9.30), however, we should rewrite these determinants – which now depend on the full metric $g_{\mu\nu}$ – in terms of σ and $\tilde{g}_{\mu\nu}$. To do this requires an understanding of how determinants transform under Weyl rescalings of the metric. This is a standard problem in field theory, and Mazur has shown that it can be solved by heat kernel techniques [193]. The outcome is an expression

$$[dg] = [d\sigma][d\xi][d\tilde{g}] e^{i\Gamma[\sigma,\tilde{g}]} \, \text{det}'|\tilde{L}^\dagger \tilde{L}|^{1/2} \frac{\text{det}|\langle \tilde{\chi}^{(\alpha)}, \tilde{\Psi}_{(\beta)} \rangle_{\tilde{g}}|}{\text{det}|\langle \tilde{\Psi}_{(\alpha)}, \tilde{\Psi}_{(\beta)} \rangle_{\tilde{g}}|^{1/2}}, \qquad (9.37)$$

where $\tilde{L}$, $\tilde{\chi}^{(\alpha)}$, and $\tilde{\Psi}_{(\alpha)}$ are defined with respect to the metric $\tilde{g}_{\mu\nu}$.

The term $\Gamma[\sigma, \tilde{g}]$ in (9.37) is a correction to the action, which can be computed explicitly in terms of the DeWitt–Seeley coefficients of the heat kernels for $L^\dagger L$ and $L L^\dagger$. If M is closed, Γ merely renormalizes Newton's constant and the cosmological constant. If M has boundaries, on the other hand, Mazur argues that a Liouville action for the conformal factor σ is induced on ∂M. This Liouville action is closely related to the boundary WZW action introduced in section 6 of chapter 2 [264], but the relationship has not yet been studied in this context.

In principle it is now possible to compute the gravitational path integral. Suppose in particular that M is a manifold for which we can choose $\tilde{R} = 0$.

Then by (9.31), the integral over σ can be easily performed:

$$\int [d\sigma] e^{iI[\tilde{g},\sigma]} = \det'|\tilde{\Delta}^{(0)}|^{-1/2}, \qquad (9.38)$$

where $\tilde{\Delta}^{(0)}$ is the Laplacian acting on scalars. The partition function then becomes

$$Z[M] = \int [d\tilde{g}] \left(\frac{\det'|\tilde{L}^\dagger \tilde{L}|}{\det'|\tilde{\Delta}^{(0)}|} \right)^{1/2} \frac{\det|\langle \tilde{\chi}^{(\alpha)}, \tilde{\Psi}_{(\beta)} \rangle_{\tilde{g}}|}{\det|\langle \tilde{\Psi}_{(\alpha)}, \tilde{\Psi}_{(\beta)} \rangle_{\tilde{g}}|^{1/2}}, \qquad (9.39)$$

where the remaining integral is over the space of diffeomorphism classes of metrics $\tilde{g}_{\mu\nu}$ such that $\tilde{R} = 0$. This is an infinite-dimensional space, whose properties are not very well understood. We shall see in the next section that a similar expression appears in the first-order formalism, but with a combination of determinants that has a known topological significance, integrated over a finite-dimensional moduli space. The partition function (9.39) exhibits tantalizing similarities to this better-understood counterpart, but the connections have not been worked out.

9.3 Path integrals and first-order quantization

We next turn to the analysis of the path integral in the first-order formalism. As in section 1 of this chapter, we could start with the canonical form of the action. If the cosmological constant vanishes, however, it turns out that a covariant path integral analogous to that of section 2 may be evaluated exactly. This is an important generalization, since it will allow us to deal with more complicated spacetime topologies, in which a canonical splitting between space and time is not globally possible. We shall therefore start with the covariant form of the first-order action,

$$I = -2 \int_M e^a \left(d\omega_a + \frac{1}{2}\epsilon_{abc}\omega^b \omega^c \right), \qquad (9.40)$$

and try to evaluate the path integral

$$Z[M] = \int [de][d\omega] e^{iI[e,\omega]} \qquad (9.41)$$

by standard methods of quantum field theory.

Let us begin with a heuristic argument. It is evident that the integral over the frame e^a in (9.41) will give a delta functional, leaving an integral over ω of the form

$$\int [d\omega]\delta [d\omega_a + \frac{1}{2}\epsilon_{abc}\omega^b \omega^c]. \qquad (9.42)$$

The spatial components of this delta functional restrict the integral to flat connections $\omega_i{}^a$ on time-slices Σ, while the time component tells us that

$$\partial_t \omega_i{}^a = \partial_i \omega_t{}^a + \epsilon^{abc} \omega_{ib} \omega_{tc}. \tag{9.43}$$

Comparing to equation (2.66), we can recognize the right-hand side of (9.43) as an $SO(2,1)$ transformation of $\omega_i{}^a$ with gauge parameter $\omega_t{}^a$. Equation (9.43) thus describes the continuous evolution of an initial flat connection by gauge transformations. The path integral then instructs us to integrate over the parameter $\omega_t{}^a$, obtaining a nonzero contribution whenever the final value of $\omega_i{}^a$ is equivalent to the initial value under some $SO(2,1)$ transformation.

This is precisely the behavior we expect from covariant canonical quantization. By exactly specifying the initial value of a flat spin connection ω, we are selecting an initial wave function that is a delta function at some value of the holonomies – in the notation of chapter 6,

$$\psi_i(\mu_\alpha) = \delta^{6g-6}(\mu_\alpha - \bar{\mu}_\alpha), \tag{9.44}$$

where $\bar{\mu}_\alpha$ are the $SO(2,1)$ holonomies on Σ_i. Similarly, the final wave function will be a delta function at some value $\bar{\mu}'_\alpha$ of the holonomies. Two such wave functions will clearly have a nonzero overlap only if the initial and final holonomies are equal. But the holonomy determines a flat spin connection uniquely up to gauge transformations, so we will obtain a nonzero transition amplitude only when the initial and final spin connections are gauge-equivalent.

To make this argument more rigorous, let us now evaluate the path integral (9.41) carefully, taking into account the need to gauge-fix the symmetries of the action [288]. To make the analysis as clear as possible, we will start with the case of a closed manifold M, avoiding the complications that arise from boundary conditions.

We shall compute the path integral using the background field method: we fix a classical extremum $(\bar{e}, \bar{\omega})$ of the action and write

$$e^a = \bar{e}^a + \alpha^a, \qquad \omega^a = \bar{\omega}^a + \beta^a. \tag{9.45}$$

It is helpful to forget for a moment that we are dealing with gravity, and to treat (9.40) as the action of a gauge theory, invariant under the transformations

$$\delta e^a = d\rho^a + \epsilon^{abc} \omega_b \rho_c + \epsilon^{abc} \omega_b \tau_c$$
$$\delta \omega^a = d\tau^a + \epsilon^{abc} \omega_b \tau_c \tag{9.46}$$

of chapter 2. A standard gauge choice is the Landau gauge,

$$\bar{g}^{\mu\nu} \bar{\nabla}_\mu \alpha_\nu{}^a = \bar{g}^{\mu\nu} \bar{\nabla}_\mu \beta_\nu{}^a = 0, \tag{9.47}$$

where $\bar{\nabla}$ is the covariant derivative with respect to some arbitrary background metric $\bar{g}$. The introduction of such a background metric is necessary for gauge-fixing; if the theory has no anomalies the final path integral will be independent of the choice of $\bar{g}$, but this will have to be checked at the end of the calculation. Note that the gauge conditions (9.47) can be rewritten in terms of the Hodge star operator $*$ for $\bar{g}$ as

$$d * e^a = d * \omega^a = 0, \tag{9.48}$$

where d is the exterior derivative and $\bar{g}$ appears only in the definition of the Hodge dual.

As Witten has pointed out [288], a more attractive form of gauge-fixing can be found if we replace d with its gauge-covariant generalization,

$$D_{\bar{\omega}} * \alpha^a = D_{\bar{\omega}} * \beta^a = 0 \tag{9.49}$$

where $D_{\bar{\omega}}$ is the gauge-covariant exterior derivative

$$D_{\bar{\omega}}\gamma^a = d\gamma^a + \epsilon^{abc}\bar{\omega}_b\gamma_c. \tag{9.50}$$

Note that if $\bar{\omega}$ is a flat connection, then $D_{\bar{\omega}}^2 = 0$. As usual, we incorporate the gauge conditions by adding Lagrange multiplier terms to the action,

$$I_{gauge} = 2\int_M (u_a D_{\bar{\omega}} * \alpha^a + v_a D_{\bar{\omega}} * \beta^a). \tag{9.51}$$

The conditions (9.49) lead to a standard Faddeev–Popov determinant,

$$\Delta_{FP} = |\det \Delta_{\bar{\omega}}^{(0)}|^2, \tag{9.52}$$

where $\Delta_{\bar{\omega}}^{(n)}$ denotes the Laplacian

$$\Delta_{\bar{\omega}} = D_{\bar{\omega}} * D_{\bar{\omega}} * + * D_{\bar{\omega}} * D_{\bar{\omega}} \tag{9.53}$$

acting on n-forms. Our gauge-fixed path integral is thus

$$Z_M = \int [d\alpha][d\beta][du][dv]|\det \Delta_{\bar{\omega}}^{(0)}|^2 \exp\{iI_{total}\} \tag{9.54}$$

where the gauge-fixed action may be written as

$$I_{total} = -2\int_M \left[\alpha^a \left(D_{\bar{\omega}}\beta_a + \frac{1}{2}\epsilon_{abc}\beta^b\beta^c + *D_{\bar{\omega}}u_a \right) \right.$$
$$\left. + \frac{1}{2}\epsilon_{abc}\bar{e}^a\beta^b\beta^c - v^a * D_{\bar{\omega}} * \beta_a \right]. \tag{9.55}$$

This action, with added ghost terms to account for the determinant (9.52), may also be obtained through BRST methods, and it has been studied in the more general context of topological BF theories [38, 39, 121, 133, 213].

Despite its rather complicated appearance, the path integral (9.54) can be evaluated exactly. It is easy to see how this works in the absence of zero-modes. First, the action (9.55) is linear in v and α, so the integration over these fields can be carried out directly, giving delta functionals

$$\delta\left[*D_{\bar{\omega}}\beta_a + \frac{1}{2}\epsilon_{abc}*(\beta^b\beta^c) + D_{\bar{\omega}}u_a\right]\delta\left[D_{\bar{\omega}}*\beta_a\right]$$

$$= \delta\left[L_-(\beta + *u)_a + \frac{1}{2}\epsilon_{abc}*(\beta^b\beta^c)\right], \quad (9.56)$$

where $L_- = (*D_{\bar{\omega}} + D_{\bar{\omega}}*)_{odd}$ maps the space of odd-dimensional forms to itself,

$$L_-(\beta + *u) = (*D_{\bar{\omega}}\beta + D_{\bar{\omega}}u) + D_{\bar{\omega}}*\beta. \quad (9.57)$$

The remaining integrations can now be carried out by using the identity (8.9). The arguments of the delta functionals (9.56) vanish when

$$D_{\bar{\omega}}\beta_a + \frac{1}{2}\epsilon_{abc}\beta^b\beta^c + *D_{\bar{\omega}}u_a = 0 = D_{\bar{\omega}}*\beta_a. \quad (9.58)$$

Clearly, one solution of these equations is $u = \beta = 0$. If this is the only solution, the connection $\bar{\omega}$ is said to be isolated and irreducible. In that case, the effect of the delta functionals (9.56) is to set ω equal to $\bar{\omega}$ in the integrand and, from (8.9), to provide a factor of $|\det L_-|^{-1}$ in the path integral.

Combining this factor with the Faddeev–Popov determinant (9.52), we obtain

$$Z_M = \frac{|\det\Delta_{\bar{\omega}}^{(0)}|^2}{|\det L_-|}. \quad (9.59)$$

This expression can be simplified by noting that

$$(L_-)^2(\beta + *u) = (*D_{\bar{\omega}}*D_{\bar{\omega}} + D_{\bar{\omega}}*D_{\bar{\omega}}*)\beta + D_{\bar{\omega}}*D_{\bar{\omega}}u$$

$$= \Delta_{\bar{\omega}}^{(1)}\beta + \Delta_{\bar{\omega}}^{(3)}*u, \quad (9.60)$$

so

$$|\det L_-|^2 = |\det\Delta_{\bar{\omega}}^{(1)}||\det\Delta_{\bar{\omega}}^{(3)}|. \quad (9.61)$$

Moreover, the Hodge $*$ operation determines a duality between n-forms and $(3 - n)$-forms; in particular, the eigenvalues of the Laplacians $\Delta_{\bar{\omega}}^{(n)}$

and $\Delta_{\bar\omega}^{(3-n)}$ are equal. Equation (9.59) can thus be written as

$$Z_M = \frac{|\det \Delta_{\bar\omega}^{(3)}|^{3/2} |\det \Delta_{\bar\omega}^{(1)}|^{1/2}}{|\det \Delta_{\bar\omega}^{(2)}|} = T[\bar\omega]. \qquad (9.62)$$

The quantity $T[\bar\omega]$ is known to topologists as the Ray–Singer torsion, or analytic torsion, of the flat connection $\bar\omega$ [227]. Its importance in physics was first pointed out by Schwarz in 1978 [237]. It can be shown that $T[\bar\omega]$ is a topological invariant, independent of the metric $\bar g$ used to define the Hodge dual. Our theory is thus anomaly-free: although the gauge-fixing required an auxiliary metric, the final result does not depend on this choice. Moreover, although the definition (9.62) involves complicated determinants of differential operators, in practice the Ray–Singer torsion can be computed for many manifolds. This is possible because $T[\bar\omega]$ is equal to a quantity known as the Reidemeister torsion, a combinatorial invariant that depends only on the cell decomposition of M and the holonomies of $\bar\omega$. Reference [66] contains an example of the computation of this invariant in the context of (2+1)-dimensional gravity, and gives an explicit result for the transition amplitude (9.62) for a particular topology-changing process.

Let us next loosen the assumption that the connection $\bar\omega$ is isolated. Suppose that equation (9.58) has infinitesimal solutions with $\beta \neq 0$. For simplicity, we shall retain the condition $u = 0$. (See reference [66] for a discussion of the case $u \neq 0$, which occurs when the connection is reducible, that is, when it is left invariant by some infinitesimal gauge transformation.) It is easy to check that $\bar\omega + \epsilon\beta$ is itself a flat $SO(2,1)$ connection; in fact, we have a whole moduli space of flat connections on M. Moreover, an infinitesimal solution of (9.58) is also a zero-mode of $\Delta_{\bar\omega}^{(1)}$, so the analytic torsion (9.62) is identically zero.

The problem here is very similar to the one we encountered at the end of section 1 when we considered the effect of Killing vectors in the ADM path integral. In the presence of zero-modes, the path integral (9.41) must be split into two pieces, with the integration over zero-modes isolated and treated separately. A careful analysis shows that (9.62) should be modified to take the form

$$Z_M = \int_{T^*\mathcal{N}} d\tilde\omega\, d\tilde e\, T[\tilde\omega],$$

$$T[\tilde\omega] = \frac{|\det' \Delta_{\tilde\omega}^{(3)}|^{3/2} |\det' \Delta_{\tilde\omega}^{(1)}|^{1/2}}{|\det' \Delta_{\tilde\omega}^{(2)}|} \qquad (9.63)$$

where the $\tilde\omega$ integral is over the moduli space $\mathcal{N}$ of flat $SO(2,1)$ connections on M, and the $\tilde e$ integral is over zero-modes of the e equation of motion (2.62) evaluated at $\tilde{\omega}$, or equivalently over the tangent space of $\mathcal{N}$.

The integrand (9.63) is once again a topological invariant, independent of any choices of background metric. Note, however, that the torsion $T[\tilde{\omega}]$ does not depend on e. Moreover, if $\tilde{e}$ is a zero-mode, so is any constant multiple of $\tilde{e}$. The $\tilde{e}$ integral thus diverges if any zero-modes are present.[§] This is potentially a serious problem: there is no known cut-off of the integral that does not destroy its purely topological character. Witten has argued that the divergences in (9.63) occur for precisely those topologies for which classical solutions of the field equations can occur, and that the divergence at large e represents the emergence of classical spacetime from the quantum theory [288]. Unfortunately, however, similar divergences can occur in topology-changing amplitudes in which e becomes large only around some interior 'wormhole', making the interpretation much less clear [66]. It is plausible that the introduction of matter will cut off the divergences, but no proof of this conjecture yet exists.

Let us next turn to the case in which the spacetime M has boundaries [66, 294]. In particular, suppose that M is a manifold with an 'initial' and a 'final' boundary, $\partial M \approx \Sigma_i \cup \Sigma_f$, where the individual boundary components may themselves be disconnected. The path integral (9.41) will then depend on the boundary data, and will give a transition amplitude between the initial and final geometries.

Not all such transitions are possible. We have seen that the only nonzero contributions to the path integral come from flat spin connections $\tilde{\omega}$ that interpolate between the initial and final surfaces. The trivial solution $\tilde{\omega} = 0$ always exists, but it clearly corresponds to degenerate geometries on Σ_i and Σ_f. If we demand that the initial and final surfaces be nondegenerate and spacelike, their topologies are severely restricted: Amano and Higuchi have shown that Σ_i and Σ_f must have equal Euler numbers, $\chi(\Sigma_i) = \chi(\Sigma_f)$ [3]. This is the same selection rule we found at the beginning of chapter 2, now appearing in a quantum mechanical context.

Suppose now that we have chosen a topology that admits a nonvanishing transition amplitude. As boundary data, we fix the induced connection $\omega_\parallel{}^a$ on ∂M. (Here, the subscript $\parallel$ denotes the tangential components; if I is the inclusion of ∂M into M, $\omega_\parallel$ means $I^*\omega$.) These boundary conditions are not quite sufficient to make the path integral well-defined; we also need some restrictions on the normal component of ω and on the triad e. It may be shown that a complete set of boundary conditions appropriate

---

[§] The existence of such zero-modes is a statement about cohomology: zero-modes $\tilde{e}$ are elements of $H^1(M; V_{\tilde{\omega}})$, while zero-modes $\tilde{\omega}$ are elements of $H^1(M, \partial M; V_{\tilde{\omega}})$, where $V_{\tilde{\omega}}$ is the flat bundle determined by $\tilde{\omega}$ [66].

for fixing ω at the boundary is

$$\beta_\| = 0 = {}^*D_{\bar\omega} {}^* \beta = 0$$
$$({}^*\alpha)_\| = {}^*D_{\bar\omega}\alpha_\| = 0. \tag{9.64}$$

In particular, these conditions make the Laplacians $\Delta_{\bar\omega}^{(n)}$ Hermitian operators.

The determinants in (9.62) now depend on boundary conditions. In particular, for any n-form γ one may define *relative*, or Dirichlet, boundary conditions

$$\gamma_\| = 0 = ({}^*D {}^* \gamma)_\| \tag{9.65}$$

and *absolute*, or Neumann, boundary conditions,

$$({}^*\gamma)_\| = 0 = ({}^*D\gamma)_\|. \tag{9.66}$$

It can be shown that the expression (9.63) remains correct as long as all of the determinants are defined either with relative boundary conditions or with absolute boundary conditions. (The two choices give equal values for the Ray–Singer torsion.) Further, the Ray–Singer torsion may once again be shown to equal the Reidemeister torsion for a manifold with boundary, making concrete computations possible [183]. Note that although it is not explicit in the expression, the amplitude (9.63) depends on the boundary values $\bar\omega^a$. The dependence arises through the range of integration: the integral in (9.63) is to be interpreted as an integral over solutions of the classical equations of motion that agree with the specified values of ω at the boundary.

Finally, let us return to the simple spacetimes $M = [0,1] \times \Sigma$ with which we began this chapter. A manifold $[0,1] \times \Sigma$ is essentially topologically equivalent to the surface Σ, and since the Ray–Singer torsion is a topological invariant, it should be possible to evaluate the expression (9.62) directly on Σ. But there is a rather easy theorem that the Ray–Singer torsion of an even-dimensional closed manifold is always unity. We have thus confirmed our original heuristic argument – the path integral for topologies $[0,1] \times \Sigma$ gives a nonzero transition amplitude if and only if the $SO(2,1)$ holonomies on Σ_i and Σ_f coincide.

9.4 Topological field theory

The specific form (9.63) of the transition amplitude for (2+1)-dimensional gravity depends on the detailed structure of the action. Many of its properties, however, can be described in the much more general language of topological quantum field theory, as axiomatized by Atiyah and Segal [17, 238].

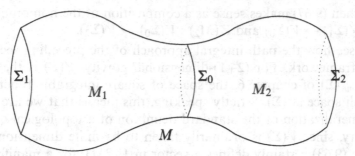

Fig. 9.1. The manifold M is formed by attaching M_1 and M_2 along a common boundary Σ_0. An axiom of topological field theory is that amplitudes 'sew' properly under such an operation.

A three-dimensional topological field theory is a theory that assigns a complex vector space $V(\Sigma)$ to each surface Σ and a vector $Z(M) \in V(\partial M)$ to each compact oriented three-manifold M, with the following properties:

1. The theory is diffeomorphism-invariant: if Σ_1 and Σ_2 are diffeomorphic, then $V(\Sigma_1) = V(\Sigma_2)$, and if M_1 and M_2 are diffeomorphic, then $Z(M_1) = Z(M_2)$.

2. $V(\overline{\Sigma}) = V(\Sigma)^*$, where $\overline{\Sigma}$ denotes Σ with reversed orientation and $V(\Sigma)^*$ is the dual space of $V(\Sigma)$.

3. For a disjoint union of two surfaces Σ_1 and Σ_2, $V(\Sigma_1 \cup \Sigma_2) = V(\Sigma_1) \otimes V(\Sigma_2)$.

4. $V(\emptyset) = \mathbf{C}$.

5. $Z([0,1] \times \Sigma)$ is the identity endomorphism of $V(\Sigma)$.

6. Suppose that $\partial M_1 = \Sigma_1 \cup \Sigma_0$, and that $\partial M_2 = \Sigma_2 \cup \Sigma_0$, and let $M = M_1 \cup_{\Sigma_0} M_2$ be the manifold formed by joining M_1 and M_2 along their common boundary Σ_0, as in figure 9.1. Then

$$Z(M) = Z(M_2) \circ Z(M_1) \in Hom(V(\Sigma_1), V(\Sigma_2)).$$

$$(9.67)$$

The last two properties require a bit of explanation. If a manifold M has two boundary components, $\partial M = \Sigma_1 \cup \Sigma_2$, then $Z(M)$ is an element of $V(\Sigma_1) \otimes V(\Sigma_2)$. But such an element can also be viewed as a homomorphism from $V(\Sigma_1)$ to $V(\Sigma_2)^* = V(\overline{\Sigma}_2)$. In particular, if $\Sigma_1 \approx \overline{\Sigma}_2$, then $Z(M)$ may be interpreted as an endomorphism of $V(\Sigma_1)$, as required by property 5. Moreover, if the orientations of the boundaries in figure 9.1 are taken to be those induced from an orientation of M_1

and M_2, then (9.67) makes sense as a composition of the homomorphisms $Z(M_1) : V(\Sigma_1) \to V(\Sigma_0)$ and $Z(M_2) : V(\Sigma_0) \to V(\Sigma_2)$.

Let us see how the path integral approach of the preceding section fits into this framework. For (2+1)-dimensional gravity, $V(\Sigma)$ is the Hilbert space $\mathcal{H}_{conn}(\Sigma)$ of chapter 6, the space of square-integrable functions on the moduli space $\mathcal{N}(\Sigma)$. Strictly speaking, this means that we are dealing with a generalization of the standard definition of a topological quantum field theory, since $V(\Sigma)$ is ordinarily taken to be finite dimensional. The amplitude (9.63) certainly defines a vector in $V(\partial M)$: for a manifold with boundary, Z_M should be viewed as a functional of the boundary data, and for each component Σ of the boundary, those data are given by a flat spin connection $\omega \in \mathcal{N}(\Sigma)$. The analytic torsion is a topological invariant, so axiom 1 is satisfied, and axioms 2–4 are fairly trivial. Note that axiom 4 is consistent with the preceding section: if M is closed, so $\partial M = \emptyset$, then $Z_M \in \mathbf{C}$. Axiom 5 is nontrivial, but we have shown it to be true: the transition amplitude for the manifold $[0, 1] \times \Sigma$ is the delta function $\delta(\mu_f - \mu_i)$.

Axiom 6 is more subtle. Physically, it is a condition on the composition of amplitudes. The intermediate surface Σ_0 in figure 9.1 carries a space of states $V(\Sigma_0)$, and the composition (9.67) represents a sum over a complete set of intermediate states. The requirement that such a composition law exist is essentially the 'sewing' condition (2.89), and the question is whether this condition is satisfied by the amplitudes (9.63).

The general answer to this question is not known, although there are results for the composition of Ray–Singer torsions that could point the way towards a proof. The Turaev–Viro model, a lattice formulation of three-dimensional quantum gravity described in chapter 11, is known to satisfy the correct sewing condition [260]. The Turaev–Viro model represents Euclidean quantum gravity with a positive cosmological constant, however, and unlike Lorentzian gravity, this model is known to have a finite-dimensional Hilbert space. The sewing property is usually assumed to hold for Lorentzian gravity as well, but at the time of this writing no direct proof is known.

Axiom 5 also has an interesting physical interpretation [27]. Consider first the application of axiom 6 to two copies of the manifold $[0, 1] \times \Sigma$. Clearly, the union $([0, 1] \times \Sigma) \cup_\Sigma ([0, 1] \times \Sigma)$ is diffeomorphic to $[0, 1] \times \Sigma$, and hence

$$Z([0, 1] \times \Sigma)^2 = Z([0, 1] \times \Sigma). \tag{9.68}$$

The propagator $Z([0, 1] \times \Sigma)$ is thus a projection operator. Next, consider an arbitrary manifold M with a boundary component diffeomorphic to Σ.

Again by (9.67),

$$Z([0,1] \times \Sigma) \circ Z(M) = Z(M \cup_\Sigma ([0,1] \times \Sigma)). \tag{9.69}$$

But $M \cup_\Sigma ([0,1] \times \Sigma)$ is diffeomorphic to M, so (9.69) implies that

$$Z([0,1] \times \Sigma) \circ Z(M) = Z(M). \tag{9.70}$$

The relationship (9.70) may be interpreted as an abstract form of the Hamiltonian constraint. The effect of sewing a 'collar' $[0,1] \times \Sigma$ onto M is to move the boundary ∂M by a diffeomorphism with a component transverse to the boundary. But such a deformation is precisely the transformation generated by the constraint $\mathcal{H}$ of chapter 2. Seen in this light, axiom 5 is the statement that $V(\Sigma)$ is the physical Hilbert space, the space of states that are invariant under the transformations generated by the Hamiltonian constraint.

If the sewing relationship (9.67) is assumed to hold, it provides a powerful method for computing amplitudes in (2+1)-dimensional quantum gravity. Consider, for example, a connected sum $M \# N$ of two three-manifolds, as described in appendix A. By definition, this new manifold is formed by cutting three-balls B^3 out of M and N and joining the resulting manifolds $M \backslash B^3$ and $N \backslash B^3$ along their new S^2 boundaries. Hence by (9.67),

$$Z(M \# N) = Z(M \backslash B^3) Z(N \backslash B^3). \tag{9.71}$$

Since the two-sphere S^2 has only one state, the composition in (9.67) is reduced to a product in (9.71). Furthermore,

$$Z(M) = Z((M \backslash B^3) \cup B^3) = Z(M \backslash B^3) Z(B^3)$$
$$Z(N) = Z((N \backslash B^3) \cup B^3) = Z(N \backslash B^3) Z(B^3), \tag{9.72}$$

and

$$Z(S^3) = Z(B^3 \cup B^3) = Z(B^3)^2, \tag{9.73}$$

since a three-sphere can be represented as two balls identified along their boundaries. Combining these results, we see that

$$Z(M \# N) = \frac{Z(M) Z(N)}{Z(S^3)}. \tag{9.74}$$

We can thus determine amplitudes for topologically complicated manifolds in terms of amplitudes of simpler constituents. In Chern–Simons theory, where a similar result holds, one can also obtain partition functions by surgery, that is, by cutting a solid torus out of a manifold M and sewing it back in a twisted fashion to obtain a new manifold M'. It is possible that such methods could be applied to (2+1)-dimensional gravity, although the situation is complicated by the fact that the gravitational Hilbert space is infinite dimensional.

One important question about this approach remains very poorly understood: we know very little about the role of the large diffeomorphisms, those diffeomorphisms that cannot be continuously deformed to the identity. In the composition (9.74), $Z(M)$ and $Z(N)$ are obtained from path integrals in which the small diffeomorphisms are factored out or gauge-fixed, but the large diffeomorphisms are not. Incorporating large diffeomorphisms into the formalism seems very difficult, since the mapping class group does not behave simply under 'sewing' [194]; for example, the mapping class group of $M\#N$ is typically very different from the mapping class groups of M and N. In reference [67], the partition function for a sphere with handles, $M = S^3 \# (S^2 \times S^1) \# \ldots \# (S^2 \times S^1)$, was evaluated, and it was shown that the result changes drastically if one factors out the large diffeomorphisms. A similar phenomenon occurs in string theory, where the proper treatment of the large diffeomorphisms leads to a string field theory that is quite different from the theory one would naively expect [299]. The corresponding treatment of gravity remains an open problem.

10

Euclidean path integrals and quantum cosmology

The first-order path integral formalism of the preceding chapter allows us to compute a large number of interesting topology-changing amplitudes, in which the universe tunnels from one spatial topology to another. It does not, however, help much with one of the principle issues of quantum cosmology, the problem of describing the birth of a universe from 'nothing'.

In the Hartle–Hawking approach to cosmology, the universe as a whole is conjectured to have appeared as a quantum fluctuation, and the relevant 'no (initial) boundary' wave function is described by a path integral for a compact manifold M with a single spatial boundary Σ (figure 10.1). In 2+1 dimensions, it follows from the Lorentz cobordism theorem of appendix B and the selection rules of page 157 that M admits a Lorentzian metric only if the Euler characteristic $\chi(\Sigma)$ vanishes, that is, if Σ is a torus. If M is a handlebody (a 'solid torus'), it is not hard to see that any resulting spacelike metric on Σ must be degenerate, essentially because the holonomy around one circumference must vanish. The case of a more complicated three-manifold with a torus boundary has not been studied, and might prove rather interesting. It is, however, atypical.

To obtain more general results, we can imitate the common procedure in 3+1 dimensions and look at 'Euclidean' path integrals, path integrals over manifolds M with positive definite metrics.[*] Since path integrals cannot be exactly computed in 3+1 dimensions, research has largely focused on the saddle point approximation, in which path integrals are dominated by some collection of classical solutions of the Euclidean Einstein field equations. One set of classical saddle points has received particular attention, the 'real tunneling geometries', which describe transitions from

---

[*] The term 'Euclidean' is traditional in quantum gravity, but it is perhaps an unfortunate usage. To mathematicians, a Euclidean metric is not merely positive definite, but flat; a curved positive definite metric is not 'Euclidean', but 'Riemannian'. In this chapter, however, I will stick to the normal physicists' usage.

163

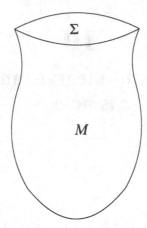

Fig. 10.1.   A manifold M with a single boundary Σ describes the birth of a universe in the Hartle–Hawking approach to quantum cosmology.

Euclidean to Lorentzian metrics [124]. Here as elsewhere in quantum gravity, the (2+1)-dimensional model permits more complete and more explicit computations, allowing us to explore implications of the sum over topologies in quantum cosmology.

10.1 Real tunneling geometries

As usual, let us begin by considering the classical gravitational configurations that are important to the quantum mechanical problem at hand. Consider two manifolds M_E and M_L with diffeomorphic boundaries $\partial M_E = \partial M_L = \Sigma$, joined along Σ into a single manifold $M = M_E \cup_\Sigma M_L$. Suppose M_E has a Euclidean metric g_E and M_L has a Lorentzian metric g_L, chosen so the metric h on Σ induced by g_E agrees with the metric induced by g_L. If the resulting geometry satisfies the Einstein field equations everywhere – perhaps with matter sources, but with no distributional source on Σ – it is said to be a real tunneling geometry. Such a geometry is an extremum of the complexified path integral, and may be thought of as providing a semiclassical description of tunneling from a Euclidean to a Lorentzian metric. In particular, M_E is closely analogous to an instanton in gauge theory, and can be used to give a semiclassical approximation for the creation of 'space' (Σ) from 'nothing', as in figure 10.1.

The requirement that the Einstein field equations be satisfied everywhere has an important implication: it means that the extrinsic curvature K_{ij} of Σ must vanish. This is most easily seen by computing K_{ij} from g_E and g_L. It is fairly easy to show that

$$K_{ij}^{(E)} = -iK_{ij}^{(L)}, \tag{10.1}$$

while the field equations require K_{ij} to be continuous [117, 124]. Hence Σ must be a surface of vanishing extrinsic curvature, that is, a totally geodesic surface.

From the Hamiltonian constraint (2.13), the vanishing of K_{ij} on Σ means that the curvature $^{(2)}R$ of Σ must satisfy

$$^{(2)}R = 2\Lambda \tag{10.2}$$

in the absence of matter. For $\Lambda > 0$, this is only possible if Σ is a two-sphere, and for $\Lambda = 0$ it requires that Σ be a torus. Moreover, matter obeying the positive energy condition can only increase the right-hand side of (10.2). To obtain more complicated spatial topologies, we must therefore require that Λ be negative.

As we have done in the rest of this book, we shall assume for simplicity that no matter is present. The Einstein field equations in three dimensions then require that g_E and g_L be metrics of constant negative curvature. In particular, M_E must be a closed hyperbolic manifold with a totally geodesic boundary. To study M_E, it is convenient to consider the 'double', denoted by $2M_E$, formed by joining two copies of M_E along Σ. The vanishing of K_{ij} on Σ guarantees that the metric on $2M_E$ is smooth across Σ, and $2M_E$ is thus a compact hyperbolic manifold. Such a manifold admits an orientation-reversing isometry θ, a 'reflection' across Σ that maps points in each copy of M_E to the corresponding points in the other copy and leaves points on Σ fixed. The original manifold M_E can be reconstructed from its double as the quotient space $2M_E/\theta$.

This construction allows us to apply the tools developed in chapter 4 for the description of geometric structures. In particular, the compact hyperbolic manifold $2M_E$ is determined uniquely by its holonomies, and can be written as a quotient of hyperbolic three-space by a discrete group Γ of isometries,

$$2M_E = \mathbf{H}^3/\Gamma. \tag{10.3}$$

Using these and more sophisticated methods, mathematicians have collected a great deal of information about such hyperbolic manifolds,[†] much of which is useful for physicists working on quantum gravity.

10.2 The Hartle–Hawking wave function

In the Hartle–Hawking approach to quantum cosmology, the wave function of the universe is expressed as a Euclidean path integral over metrics and matter fields on a manifold M with a single boundary component Σ

[†] A standard reference is a set of unpublished but widely circulated notes by Thurston, reference [256].

[140]. Such a path integral depends on the induced metric h_{ij} and the matter configuration $\varphi|_{\partial M}$ on Σ, thus determining a functional

$$\Psi[h, \varphi|_{\partial M}] = \sum_M \int [dg][d\varphi] \, \exp\{-I_E[g, \varphi]\}. \qquad (10.4)$$

The summation represents a sum over topologies of M; all manifolds with a given boundary Σ are assumed to contribute. The Hartle–Hawking wave function Ψ is to be interpreted as an amplitude for finding a universe characterized by h and $\varphi|_{\partial M}$. This approach finesses the question of initial conditions for the universe by simply omitting an initial boundary, and it postpones the question of the nature of time in quantum gravity: information about time is hidden in the boundary geometry h, but the path integral can be formulated without making a choice of time explicit.

In three spacetime dimensions, the Euclidean gravitational action for a manifold M with boundary Σ is

$$I_E[g_E] = -\frac{1}{16\pi G} \int_M d^3x \, \sqrt{g_E} \, (R[g_E] - 2\Lambda) - \frac{1}{8\pi G} \int_\Sigma d^2x \, \sqrt{h} K, \qquad (10.5)$$

where the boundary term – the trace of the extrinsic curvature of the boundary – is the one appropriate for fixing the metric on Σ [125]. From the arguments of the preceding section, we expect the path integral to be dominated by real tunneling geometries, for which the boundary term vanishes. The classical action is then

$$\bar{I}_E = \frac{1}{4\pi G|\Lambda|^{1/2}} Vol(M), \qquad (10.6)$$

where $Vol(M)$ is the volume of M with the metric rescaled to constant curvature -1.

Classically, real tunneling geometries describe the transition from Euclidean to Lorentzian signature metrics. In the quantum theory, however, there is an important subtlety: if we specify the extrinsic curvature K_{ij} on Σ, as required for a smooth continuation to Lorentzian signature, then we cannot also freely specify the induced metric h_{ij}, which is canonically conjugate to K_{ij}. The wave function (10.4) is a functional of the boundary data, and varies as we vary h_{ij}. But if Σ is required to be totally geodesic, the induced metric h_{ij} is *unique* up to diffeomorphisms. A given real tunneling geometry can thus contribute to the wave function (10.4) at only a single value $\hat{h}_{ij}$ of its argument, and cannot be used to obtain even an approximation of the full wave function.

This difficulty comes in part from the fact that we have looked only at the contribution of a single topology to the sum in (10.4). Real tunneling geometries on topologically distinct manifolds M_1 and M_2 will induce different metrics $\hat{h}_{ij}^{(1)}$ and $\hat{h}_{ij}^{(2)}$ on Σ; in fact, if all topologies are taken into account, these contributions will fill out a dense set of the moduli space of Σ [115]. Moreover, if we fix a manifold M and allow the metric and extrinsic curvature on Σ to vary, it may be shown that the metric $\hat{h}_{ij}$ induced by the real tunneling geometry on M is a local extremum of the Hartle–Hawking wave function [57]. It is likely, although not yet proven, that this extremum is a maximum among metrics at the fixed York time $TrK = 0$.

Real tunneling geometries on a fixed manifold thus determine the extrema of the Hartle–Hawking wave function, the spatial geometries most likely to be reached by a tunneling process from 'nothing', and the sum over topologies gives us information about a large portion of moduli space. The interplay between extrema of $\Psi[h]$ for a single topology and the summation over topologies is a subtle one, however, which will be discussed further in the next section.

Since we are now working with a model in which the cosmological constant is nonzero, the exact computations of the preceding chapter are no longer valid. We must evaluate the path integral (10.4) perturbatively, expanding around the classical extrema g_E in a saddle point approximation. For a given real tunneling geometry on a manifold M, the lowest order term is simply the exponential of the classical action,

$$\exp\left\{-\bar{I}_E\right\} = \exp\left\{-\frac{1}{4\pi G|\Lambda|^{1/2}}Vol(M)\right\}, \tag{10.7}$$

and we are instructed to sum this quantity over topologies.

The quantity $Vol(M)$ is a topological invariant of M. Moreover, the spectrum of volumes is discrete, although it has accumulation points. As a first guess – to which we shall return in the next section – we might therefore expect the sum over topologies to be dominated by those manifolds for which (10.7) is greatest, that is, for which $Vol(M)$ is smallest [117, 118].

If Σ is a genus two surface, the real tunneling geometries of minimal volume are known explicitly [169]. There are eight contributions from topologically distinct manifolds with equal volumes, each of which can be constructed by gluing two hyperbolic tetrahedra along their faces. This means that we can write an explicit expression for the moduli of Σ at which the Hartle–Hawking wave function is maximized. Similar results exist for higher genus spaces. The minimum volume of a hyperbolic manifold with a totally geodesic boundary of genus g is also known; it is approximately proportional to g, a fact which could be interpreted as implying that low genus spaces are 'more probable'.

By taking advantage of the connection between three-dimensional gravity and Chern–Simons theory, we can also compute the first quantum correction to the amplitude (10.7). This correction consists of a collection of determinants similar to the determinants of equation (9.62), coming from gauge-fixing and from quadratic fluctuations around the extremal value of the metric:

$$\Psi[h] \sim \sum_M \Delta_M \exp\left\{-\frac{1}{4\pi G|\Lambda|^{1/2}}Vol(M)\right\}. \qquad (10.8)$$

To evaluate the prefactor Δ_M, note that three-dimensional Euclidean gravity may be expressed as a Chern–Simons theory, in this case for the gauge group $SL(2,\mathbf{C})$. As in chapter 2, the $SL(2,\mathbf{C})$ connection is

$$A_\mu{}^a = \frac{1}{2}\epsilon^{abc}\omega_{\mu bc} + i|\Lambda|^{1/2}e_\mu{}^a. \qquad (10.9)$$

Moreover, as in chapter 4, A is a flat connection that is completely determined by the geometric structure of M, and it can be computed from the uniformizing group Γ of equation (10.3). Standard results from Chern–Simons theory then tell us that the prefactor Δ_M is related to the Ray–Singer torsion of this connection [289],

$$\Delta_M = T^{1/2}(M,A), \qquad (10.10)$$

where $T(M,A)$ is defined as in equation (9.62). For a given M, this quantity is calculable, and as in chapter 9, its computation can be reduced to the combinatorial problem of determining the associated Reidemeister torsion. For the case of genus two, for example, the holonomies of the minimum-volume real tunneling geometries have been computed in reference [116], and it should be possible to use those results to determine the Ray–Singer torsion.

10.3 The sum over topologies

As we noted above, the locations of the peaks of the Hartle–Hawking wave function – the 'most probable universes' in Hawking's quantum cosmology – depend on a delicate interplay between two factors. For a three-manifold M of a given topology, the wave function $\Psi[h]$ is extremized when $K_{ij} = 0$ on the boundary. The dominant contribution comes from the corresponding real tunneling geometry, and is largest for the geometries with the smallest volumes. On the other hand, we must also sum over an infinite number of manifolds with different topologies, and it is possible that the 'entropy' might dominate: if enough manifolds contribute at a given boundary metric $\hat{h}_{ij}$, their number could overcome the exponential suppression in equation (10.8).

In general, the balance between volume and entropy is an open question, but a particular example is known in which the entropy dominates [54]. Neumann and Reid have found an infinite set of hyperbolic manifolds $\widetilde{M}_{(p,q)}$, where p and q are relatively prime integers, with the following characteristics [210, 211]:

1. each of the $\widetilde{M}_{(p,q)}$ has a single totally geodesic boundary, with a fixed hyperbolic metric $\hat{h}_\infty$ that is independent of p and q;

2. the volumes of the $\widetilde{M}_{(p,q)}$ are bounded above by a finite number $Vol(\widetilde{M}_\infty)$, and converge to $Vol(\widetilde{M}_\infty)$ as $p^2 + q^2 \to \infty$; and

3. the Ray–Singer torsions $T(\widetilde{M}_{(p,q)}, A_{(p,q)})$ in the prefactors (10.10) take on a dense set of values in the interval $(0, cT_\infty]$, where cT_∞ is a positive constant.

These properties imply that the $\widetilde{M}_{(p,q)}$ all give positive contributions to the Hartle–Hawking wave function at $h = \hat{h}_\infty$. Indeed, conditions 2 and 3 guarantee that the sum (10.8) diverges at $\hat{h}_\infty$: the volumes converge to $Vol(\widetilde{M}_\infty)$, while infinitely many of the prefactors are bounded below by some $\epsilon > 0$. The Hartle–Hawking wave function is thus infinitely peaked at $\hat{h}_\infty$, even though the Neumann–Reid manifolds do not minimize the exponent in (10.8).

The construction of the $\widetilde{M}_{(p,q)}$ is described in detail in reference [54]. It is based on a procedure called hyperbolic Dehn surgery, in which a singular cusp in an initial manifold M_∞ is 'filled in' by cutting out a neighborhood of the singularity and gluing in a solid torus. The integers p and q describe the way the torus is twisted before it is glued in. The spatial manifolds Σ generated by the Neumann–Reid construction are not very general – the simplest has genus $g = 50$ – but the existence of this divergence should serve as a warning that the sum over topologies may lead to surprising results.

There is a suggestive argument, based on the requirement of 'rigidity' in the Neumann–Reid construction, that divergences of this type may occur only at isolated points in the moduli space of Σ. In general, a procedure such as hyperbolic Dehn surgery will change the metric on the boundary Σ, thus smearing out any divergence in the sum over topologies. This did not occur in the Neumann–Reid example for a very specific reason: their boundary was realized as a covering space of a rigid surface, the two-sphere with three conical singularities. Here, 'rigid' means that the surface admits only one constant negative curvature metric, i.e., that its moduli space consists of a single point. Only a few, highly symmetric surfaces occur as covering spaces of rigid surfaces, and it is plausible that the sum over topologies will diverge only for such surfaces. If this is the case, it may be possible to give a complete description of the normalized Hartle–Hawking wave function as a sum of delta functions at isolated points in moduli space.

Three-dimensional gravity provides at least two other examples of unexpected divergences arising from the sum over topologies. In the computation of the partition function for closed three-manifolds with $\Lambda > 0$, the path integral includes contributions from an infinite number of lens spaces $L_{p,q}$, manifolds obtained from the three-sphere by suitable identifications. It is shown in reference [56] that this sum diverges, in this case because of the contribution of low-volume, topologically complicated manifolds. Another divergence occurs in the sum over 'wormholes', connected sums of copies of $S^2 \times S^1$, in (2+1)-dimensional gravity with Lorentzian signature [67].

None of these examples extends directly to 3+1 dimensions, but the (2+1)-dimensional results should at least make one cautious. The extrema of the action in 3+1 dimensions need not have constant negative curvature, making a detailed analysis much more difficult. But constant negative curvature four-manifolds are still extrema of the Einstein–Hilbert action, if not the only ones, and it has now been shown that the number of such manifolds with volume less than x grows at least factorially with x [64]. Using arguments similar to those developed here, we may again conclude that entropy almost certainly dominates action in the Hartle–Hawking wave function of the real (3+1)-dimensional universe.

11

Lattice methods

In a number of quantum field theories – quantum chromodynamics, for example – a standard approach to conceptual and computational difficulties is to discretize the theory, replacing continuous spacetime with a finite lattice. The path integral for a lattice field theory can be evaluated numerically, and insights from lattice behavior can often teach us about the continuum limit. Gravity is no exception: one of the earliest pieces of work on lattice field theory was Regge's discretization of general relativity [228], and the study of lattice methods continues to be an important component of research in quantum gravity.

Like other methods, lattice approaches to general relativity become simpler in 2+1 dimensions. Classically, a (2+1)-dimensional simplicial description of the Einstein field equations is, in a sense, exact: tetrahedra may be filled in by patches of flat spacetime, and it is only at the boundaries, where patches meet, that nontrivial dynamics can occur. This means, among other things, that the constraints of general relativity are much easier to implement. Recall that the constraints generate diffeomorphisms, and can thus be thought of as moving points, including the vertices of a lattice. In 3+1 dimensions, this causes serious difficulties. In 2+1 dimensions, however, the geometry is insensitive to the location of the vertices, so such transformations are harmless. Equivalently, the diffeomorphisms can be traded for gauge transformations in the Chern–Simons formulation of (2+1)-dimensional gravity, and these act pointwise and preserve the lattice structure. Similarly, the loop representation of chapter 7 is naturally adapted to a discrete description: as long as a lattice is fine enough to capture the full spacetime topology, the holonomies along edges of the lattice provide a natural (over)complete set of loop operators.

Lattice descriptions of (2+1)-dimensional gravity have surprising connections to a wide variety of mathematical and physical structures: topological invariants of three-manifolds and knots, category theory, quantum

171

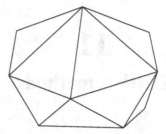

Fig. 11.1. For suitable metrics, the curvature of a triangulated surface is concentrated entirely at the vertices, and is determined by the conical deficit angle at each vertex.

groups, and topological field theory, to name a few. The subject has been active enough to merit a book of its own; indeed, much of Turaev's *Quantum Invariants of Knots and 3-Manifolds* may be read as a program to place three-dimensional lattice quantum gravity and its generalizations on a mathematically sound footing [259]. This chapter will not attempt to provide a comprehensive survey of work in this area, but should rather be approached as an outline and a guide for further reading.

11.1 Regge calculus

The lattice formulation of general relativity was first developed by Regge in the early 1960s [228]. The idea of 'Regge calculus' is most easily understood in the case of a two-dimensional triangulated surface, as illustrated in figure 11.1. The triangles in this figure are flat, and the edges carry only extrinsic curvature; the intrinsic curvature is concentrated entirely at the vertices. These vertices are conical points, characterized by deficit angles δ_i. But we know from chapter 3 that each such point contributes an amount $2\delta_i$ to the integral $\int d^2x \sqrt{g}R$. The Einstein action for a triangulated surface is thus

$$I_{Regge}^{(2)} = 2 \sum_{\text{vertices}:i} \delta_i. \tag{11.1}$$

(This is, of course, a Euclidean action, that is, an action for positive definite metrics.)

Regge showed that a similar expression holds for an n-dimensional simplicial manifold M, composed of flat n-simplices glued along their $(n-1)$-dimensional edges. The curvature is again concentrated entirely on the $(n-2)$-dimensional 'hinges' (or 'bones'), and the Euclidean Einstein action takes the form

$$I_{Regge}^{(n)} = 2 \sum_{\text{hinges}:i} \delta_i V_i, \tag{11.2}$$

where δ_i is again the deficit angle and V_i is the volume of the ith hinge. In particular, the action for three-dimensional general relativity is

$$I_{Regge} = 2 \sum_{\text{edges}:e} \delta_e \ell_e, \tag{11.3}$$

where ℓ_e is the length of the eth edge.* A similar expression exists for a simplicial manifold with a metric of Lorentzian signature, although the definition of the deficit angle is slightly more complicated [29]: the Lorentzian angle θ_e between two spacelike faces meeting at an edge E_e is essentially the boost parameter that takes the normal of one surface to the normal of the other,

$$\theta_e = \pm \cosh^{-1}(n_1 \cdot n_2), \tag{11.4}$$

and δ_e is the sum of all such angles around a given edge. The Regge action reduces to the standard Einstein–Hilbert action as edge lengths go to zero, and one can derive such standard results as the weak-field limit of Einstein gravity [139].

Regge's discretized action suggests a new path integral approach, particularly if one is interested in the Euclidean path integral of chapter 10. Consider a simplicial three-manifold M with a triangulated boundary $\Sigma_0 \cup \Sigma_1$. A state on, say, Σ_0 can be viewed as a function of the lengths $\{\ell_e(0)\}$ of edges lying on Σ_0. (This description will be refined below.) The transition amplitude between a state on Σ_0 and a state on Σ_1 may then be expressed schematically as a path integral

$$Z[\{\ell_e(0)\}, \{\ell_e(1)\}] = \int_{\substack{\text{interior} \\ \text{edges}}} d\ell_e \prod_{\text{edges}:e} \exp\{-2\delta_e \ell_e\}. \tag{11.5}$$

The path integral is thus reduced to a finite-dimensional integral, which can in principle be evaluated numerically. The appropriate measure in this integral is uncertain; it should presumable be deduced from the full three-dimensional path integral, but the role of gauge-fixing and the correct form of the Faddeev–Popov determinants are poorly understood.

Three-dimensional Regge calculus has a remarkable feature, first noted by Ponzano and Regge: the path integral (11.5) can be reexpressed in terms

---

* Many papers on Regge calculus use units $8\pi G = 1$, and thus omit the factor of 2 in (11.3). Since my convention is to take $16\pi G = 1$, expressions in this section will sometimes differ from those in the literature.

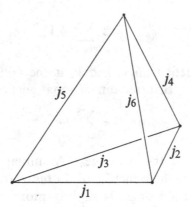

Fig. 11.2.　A coloring of a graph is an assignment of an integer or half-integer to each edge.

of Wigner–Racah $6j$-symbols, which describe the coupling of three angular momenta in quantum mechanics [221]. (See, for example, references [72] or [174] for a more detailed description of these coefficients.) Consider a tetrahedron T with a 'coloring', that is, an assignment of an integer or half-integer j_i to each edge, as illustrated in figure 11.2. With any such coloring, we can associate the $6j$-symbol

$$\left\{ \begin{array}{ccc} j_1 & j_2 & j_3 \\ j_4 & j_5 & j_6 \end{array} \right\}. \tag{11.6}$$

Note that $6j$-symbols have tetrahedral symmetry, so this assignment is unique. Moreover, since $6j$-symbols describe the composition of angular momenta, they are nonzero only when certain triangle inequalities are satisfied: $|j_1 - j_2| \le j_3 \le j_1 + j_2$, along with corresponding relations for (j_1, j_5, j_6), (j_4, j_2, j_6), and (j_4, j_5, j_3). From figure 11.2, we see that these are simply the triangle inequalities for the faces of a tetrahedron with edge lengths proportional to the j_i.

Ponzano and Regge show that for large j,

$$\exp\{\pi i \sum_{i=1}^{6} j_i\} \left\{ \begin{array}{ccc} j_1 & j_2 & j_3 \\ j_4 & j_5 & j_6 \end{array} \right\} \sim$$

$$\frac{1}{\sqrt{6\pi V}} \left[\exp\left\{ i\left(I_{Regge} + \frac{\pi}{4} \right) \right\} + \exp\left\{ -i\left(I_{Regge} + \frac{\pi}{4} \right) \right\} \right]. \tag{11.7}$$

Here, I_{Regge} is the Regge action for a tetrahedron with sides of length $\ell_i = \frac{1}{2}(j_i + \frac{1}{2})$,

$$I_{Regge} = \sum_{i=1}^{6} \theta_i (j_i + \frac{1}{2}), \tag{11.8}$$

where θ_i is the angle between the outward normals of the two faces meeting at the ith edge and V is the volume of the tetrahedron with these edge lengths.

Since the Regge action for an arbitrary simplicial manifold can be expressed as a sum of actions for individual tetrahedra, we can thus convert the integrand of (11.5) into a collection of products of $6j$-symbols. This step is somewhat delicate, thanks to the appearance of positive- and negative-frequency terms in (11.7), but we should at least obtain one term of the form $\exp\{iI_{Regge}\}$ in such a collection of products. Similarly, the integral over lengths of interior edges should, for large enough j, be equivalent to a sum over spins, although the appropriate measure must be determined and some regularization may be needed. The discretized Euclidean path integral is thus, arguably, a sum of products of $6j$-symbols.

Ponzano and Regge propose the following expression for such a regularized sum. (The expression below includes modifications introduced by Ooguri to account for boundaries [216].) Consider a manifold M with boundary ∂M, with a given triangulation Δ of ∂M. Choose a triangulation of M that agrees with the triangulation of the boundary, that is, such that each triangle in ∂M is a face of a tetrahedron in M. Label interior edges of tetrahedra by integers or half-integers x_i and exterior (boundary) edges by j_i, and for a given tetrahedron t, let $j_i(t)$ denote the spins that color its (interior and exterior) edges. Then

$$Z_\Delta[\{j_i\}] = \lim_{L \to \infty} \sum_{x_e \le L} \left(\prod_{\text{ext. edges:}i} (-1)^{2j_i} \sqrt{2j_i + 1} \prod_{\text{int. vertices}} \Lambda(L)^{-1} \right.$$

$$\left. \prod_{\text{int. edges:}\ell} (2x_\ell + 1) \prod_{\text{tetra:}t} (-1)^{\sum_{i=1}^{6} j_i(t)} \left\{ \begin{array}{ccc} j_1(t) & j_2(t) & j_3(t) \\ j_4(t) & j_5(t) & j_6(t) \end{array} \right\} \right), \quad (11.9)$$

where 'int' and 'ext' mean 'interior' and 'exterior' and

$$\Lambda(L) = \sum_{j \le L} (2j + 1)^2 \quad (11.10)$$

is a regularization factor that controls divergences in the sum over interior lengths.

With this choice of weighting, identities among $6j$-symbols may be used to show that the transition amplitude is invariant under subdivision (or 'refinement'): that is, if a tetrahedron is replaced by four tetrahedra as shown in figure 11.3, Z is unchanged. This suggests, although it does not yet prove, that we are dealing with a topological field theory, a theory that depends only on the topology of M and not on the triangulation we happen to choose. Such a property is not expected in $(3+1)$-dimensional gravity – by subdividing a simplex, we would presumably obtain finer

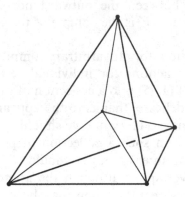

Fig. 11.3.   A subdivision of a tetrahedron is constructed by adding a vertex at its center and edges that join this new vertex to those of the original tetrahedron, yielding a set of four smaller tetrahedra.

grained information about the geometry – but it is not so surprising in 2+1 dimensions, given the global nature of the classical solutions.

11.2 The Turaev–Viro model

The regularization (11.9) is rather difficult to use, and it is hard to obtain exact results in the Ponzano–Regge model. Fortunately, Turaev and Viro have discovered an improved regularization, based on the technology of quantum groups [260]. The resulting Turaev–Viro model can be shown to be a genuine topological field theory, satisfying the axioms of the last section of chapter 9.

To construct the Turaev–Viro model, let k be a positive integer, and define

$$q = \exp\left\{\frac{2\pi i}{k+2}\right\}. \tag{11.11}$$

The quantum group $U_q(sl(2))$ ('quantum $SU(2)$') has representations labeled by integers and half-integers $j \leq k/2$, so if the spins j are replaced by their quantum group analogs, the sum in (11.9) will be automatically cut off, while the $L \to \infty$ limit will correspond to $k \to \infty$. Now let

$$[n] = \frac{q^{n/2} - q^{-n/2}}{q^{1/2} - q^{-1/2}} \tag{11.12}$$

and

$$\Lambda_q = -\frac{2(k+2)}{(q^{1/2} - q^{-1/2})^2}, \tag{11.13}$$

and denote the quantum $6j$-symbols by

$$\left\{ \begin{array}{ccc} j_1 & j_2 & j_3 \\ j_4 & j_5 & j_6 \end{array} \right\}_q \tag{11.14}$$

(see references [72] or [260] for a detailed definition). The Turaev–Viro version of the transition amplitude (11.9) is then

$$Z_\Delta^{(k)}[\{j_i\}] = \sum_{\text{admis.}\,\{x_e\}} \left(\prod_{\text{ext. edges:}i} e^{\pi i j_i}[2j_i+1]^{1/2} \prod_{\text{ext. vertices}} \Lambda_q^{-1/2} \prod_{\text{int. vertices}} \Lambda_q^{-1} \right.$$

$$\left. \prod_{\text{int. edges:}\ell} e^{2\pi i x_\ell}[2x_\ell+1] \prod_{\text{tetra:}t} (-1)^{\sum_{i=1}^{6} j_i(t)} \left\{ \begin{array}{ccc} j_1(t) & j_2(t) & j_3(t) \\ j_4(t) & j_5(t) & j_6(t) \end{array} \right\}_q \right),$$

$$\tag{11.15}$$

where a set of spins $\{x_e\}$ is admissible ('admis.') if it satisfies the appropriate triangle inequalities along with a set of inequalities coming from the representation theory of $U_q(sl(2))$ – e.g., $x_e \le k/2$. Turaev and Viro show that this expression is independent of the triangulation of the interior, even for finite k. By choosing a simple enough triangulation, one can actually compute Z for many manifolds [260, 286].

Let us now return to the problem, mentioned briefly in the preceding section, of defining the physical states of this model. Fix a surface Σ and a triangulation Δ. The spins j_i on edges E_i of Δ give the discrete analog of the spatial metric, and a natural candidate for the space of states is the vector space $V^{(k)}(\Sigma, \Delta)$ of functions $\phi_\Delta(\{j_i\})$ of the spins. For a given k this space is finite dimensional, since only finitely many colorings of (Σ, Δ) are admissible. It is still much too large, however: we have not yet imposed the discrete analogs of the momentum and Hamiltonian constraints of quantum gravity.

Discretized expressions for the constraints (2.13) and (2.14) of chapter 2 are not easy to construct. There is, however, a simpler way to impose the constraints, as we saw at the end of chapter 9. Consider the manifold $M = [0,1] \times \Sigma$, and choose a simplicial decomposition of M such that the induced triangulation Δ is identical on the boundaries $\{0\} \times \Sigma$ and $\{1\} \times \Sigma$. Select colorings $\{j_i(0)\}$ and $\{j_i(1)\}$ at the two boundaries. The Turaev–Viro amplitude (11.15) then defines a propagator

$$Z_{\Delta,\Delta}^{(k)}[\{j_i(0)\}, \{j_i(1)\}] = P_\Delta[\{j_i(0)\}, \{j_i(1)\}], \tag{11.16}$$

which in turn determines a map $\hat{P}_\Delta$ from $V^{(k)}(\Sigma, \Delta)$ to itself: a function $\phi_\Delta(\{j_i\})$ maps to

$$(\hat{P}_\Delta \phi)_\Delta(\{j_i\}) = \sum_{j_\ell(1)} P_\Delta[\{j_i(0)\}, \{j_\ell(1)\}]\phi_\Delta(\{j_\ell(1)\}). \tag{11.17}$$

It is not hard to see that this map is a projection operator, that is, $\hat{P}_\Delta^2 = \hat{P}_\Delta$. Indeed, $\hat{P}_\Delta^2$ may be obtained by attaching two copies of $[0, 1] \times \Sigma$ along a common boundary, a process that yields a manifold diffeomorphic to $[0, 1] \times \Sigma$, and it is straightforward to show that the sum over colorings of the common boundary then reproduces the Turaev–Viro amplitude (11.16). This is a particular version of the 'sewing' argument of the last section of chapter 9, and it provides one of the few cases in which this property can be proven explicitly.

As a projection operator, $\hat{P}_\Delta$ has eigenvalues 0 and 1. But we know from sections 3 and 4 of chapter 9 how this operator must act on the physical Hilbert space: the physical propagator is just a delta function. The Turaev–Viro Hilbert space $\mathcal{H}_{TV}$ may thus be taken to be the subspace of $V^{(k)}(\Sigma, \Delta)$ consisting of eigenstates of $\hat{P}_\Delta$ with eigenvalue 1.

This definition appears to depend on the triangulation of Σ, but in fact it does not. To see this, consider the manifold $[0, 1] \times \Sigma$, now with different triangulations Δ_0 and Δ_1 on the boundaries $\{0\} \times \Sigma$ and $\{1\} \times \Sigma$. The amplitude $Z^{(k)}_{\Delta_0, \Delta_1}[\{j_i(0)\}, \{j_\ell(1)\}]$ then determines a map $\hat{P}_{\Delta_0, \Delta_1}$ from $V^{(k)}(\Sigma, \Delta_1)$ to $V^{(k)}(\Sigma, \Delta_0)$. The same kind of 'sewing' argument that showed that $\hat{P}_\Delta$ was a projection operator now shows that $\hat{P}_{\Delta_0, \Delta_1} \hat{P}_{\Delta_1, \Delta_0} = \hat{P}_{\Delta_0}$, which is the identity on the Hilbert space determined by Δ_0, and that similarly $\hat{P}_{\Delta_1, \Delta_0} \hat{P}_{\Delta_0, \Delta_1} = \hat{P}_{\Delta_1}$. The map $\hat{P}_{\Delta_0, \Delta_1}$ thus determines an isomorphism between the Hilbert spaces defined with respect to the two triangulations.

Now that we have a Hilbert space $\mathcal{H}_{TV}$ and transition amplitudes (11.15), we can explore the connections between lattice quantum gravity and the continuum theory in more detail. A number of interesting results have been found over the past several years:

1. For large (but finite) k, the Turaev–Viro amplitude for the tetrahedron has a limit similar to expression (11.7), with the Regge action again appearing in the exponent [199, 200]. The effect of finite k can be computed by using identities among quantum $6j$-symbols. One finds a correction equivalent to the addition of a positive cosmological constant

$$\Lambda = \left(\frac{4\pi}{k}\right)^2 \tag{11.18}$$

   to the Regge action.

2. In the large k limit, the Turaev–Viro Hilbert space $\mathcal{H}_{TV}$ may be shown to be isomorphic to the space of gauge-invariant functions of flat $SU(2)$ connections on Σ [216, 218, 234]. This establishes a direct tie to the connection representation of chapter 6: just as

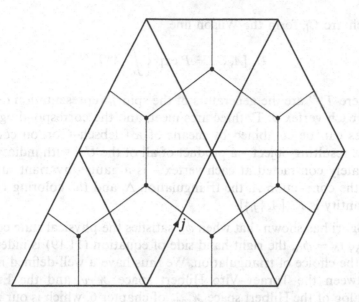

Fig. 11.4. The graph dual to a triangulated surface is formed by connecting the centers of adjacent triangles. A coloring of the triangulation determines a coloring of the dual graph; in this figure, for instance, j is associated with both an edge of a triangle (shown by an arrow) and the segment of the dual graph that crosses that edge.

(2+1)-dimensional Lorentzian gravity can be written as an $ISO(2,1)$ Chern–Simons theory with a configuration space of flat $SO(2,1)$ connections, three-dimensional Euclidean gravity can be written as an $ISU(2)$ Chern–Simons theory with a configuration space of flat $SU(2)$ connections.[†]

To construct this isomorphism, begin with a surface Σ with a fixed triangulation Δ, choose a flat $SU(2)$ connection A, and write

$$\psi[A] = \sum_{j_i} K_\Delta[A, \{j_i\}]\phi_\Delta(\{j_i\}), \qquad (11.19)$$

where ϕ_Δ is an element of the space $V^{(k)}(\Sigma, \Delta)$. The transformation matrix $K_\Delta[A, \{j_i\}]$ is defined as follows. Let Γ be the trivalent graph dual to Δ, formed by assigning one vertex to each triangle of Δ and connecting each pair of vertices separated by a single edge of Δ (see figure 11.4). Each arc C_i in Γ crosses one edge of E_i of Δ, and can be assigned a corresponding spin j_i, as illustrated in the figure. For

---

[†] The question of whether the gauge group for Euclidean gravity should be $SU(2)$ or $SO(3)$ is a subtle one; see reference [51] for a related discussion of whether the gauge group for (2+1)-dimensional gravity should be $ISO(2,1)$ or its universal cover.

---

each arc C_i, form the Wilson line

$$U_{j_i}[A, C_i] = P \exp\left\{\int_{C_i} A^a T_a^{(j_i)}\right\},\qquad(11.20)$$

where $T_a^{(j_i)}$ are the generators of the spin j_i representation of $SU(2)$. At each vertex of Γ, three arcs meet, and the corresponding Wilson lines can be combined by means of a Clebsch–Gordon coefficient. The resulting object – a product of all of the Us, with indices appropriately contracted at each vertex – is a gauge-invariant functional of the connection A, the triangulation Δ, and the coloring $\{j_i\}$. This quantity is $K_\Delta[A, \{j_i\}]$.

Ooguri has shown that when ϕ_Δ satisfies the physical state condition $(\hat{P}_\Delta\phi)_\Delta = \phi_\Delta$, the right-hand side of equation (11.19) is independent of the choice of triangulation. We thus have a well-defined mapping between the Turaev–Viro Hilbert space $\mathcal{H}_{TV}$ and the Euclidean analog of the Hilbert space $\mathcal{H}_{conn}$ of chapter 6, which is our required isomorphism.

3. The transformation (11.19) is closely related to the loop transform of chapter 7 [232].‡ There are two obvious differences: the kernel $K_\Delta[A, \{j_i\}]$ depends on a trivalent graph rather than a set of loops, and it depends on Wilson lines for arbitrary representations rather than the spin 1/2 representation alone. But standard techniques based on Penrose's binor calculus for spin networks may be used to transform $K_\Delta[A, \{j_i\}]$ into a collection of traces of products of spin 1/2 Wilson loops. Schematically, one replaces a Wilson loop in the spin j representation with a set of $2j$ loops in the spin 1/2 representation, and sums over all routings of the loops that meet at a vertex. When this is done, the result is precisely the Euclidean version of the loop transform (7.33). Note that for Euclidean signature, the difficulties in defining the loop transform discussed at the end of chapter 7 are not present.

Rovelli has argued that in the loop representation, the spins j are the eigenvalues of a length operator, as one might expect from the original Ponzano–Regge construction [232]. The length of a curve C in classical three-dimensional gravity is

$$\ell[C] = \int_C ds \left(g_{\mu\nu}\frac{dC^\mu}{ds}\frac{dC^\nu}{ds}\right)^{1/2},\qquad(11.21)$$

---

‡ See also references [72] and [167] for relationships between the Turaev–Viro model and spin networks.

and as we saw in chapter 7, the corresponding operator $\hat{\ell}[C]$ can be defined in terms of the $\hat{T}$ operators of the loop representation. It may be shown that for each edge E_i of a triangulation Δ, the state $\phi_\Delta(\{j_i\})$ is an eigenstate of the operator $\hat{\ell}[E_i]$ with eigenvalue j_i. This is not quite an assertion about physical states and gauge-invariant operators – it depends on the choice of triangulation, while the physical Hilbert space does not – but it is nevertheless suggestive.

4. Let M be a closed three-manifold. Then for finite k, the Turaev–Viro amplitude (11.15), which should now be interpreted as a partition function, is equal to the absolute square of the partition function for an $SU(2)$ Chern–Simons theory with coupling constant k,

$$Z_{TV} = |Z_{CS}|^2 \qquad (11.22)$$

[224, 258]. This again establishes an equivalence with Euclidean gravity in first-order form. Indeed, we saw in chapter 2 that the first-order action for (2+1)-dimensional Lorentzian gravity could be written as a Chern–Simons action, with a gauge group that depended on the sign of the cosmological constant. The same is true for the Euclidean theory: the gravitational action with positive cosmological constant Λ can be written as

$$I = I_{CS}[A^+] - I_{CS}[A^-], \qquad (11.23)$$

where I_{CS} is the Chern–Simons action (2.71) for the gauge group $SU(2)$, with coupling constant

$$k = \frac{4\pi}{\sqrt{\Lambda}} \qquad (11.24)$$

and gauge fields

$$A^\pm = \omega \pm \sqrt{\Lambda}e. \qquad (11.25)$$

Equation (11.24) agrees with the expression (11.18) for the cosmological constant as determined from the asymptotics of quantum $6j$-symbols. Moreover, the path integral for the action (11.23) is, heuristically,

$$Z_{CS} = \int [dA^+][dA^-]e^{i(I[A^+]-I[A^-])} = \left| \int [dA^+]e^{iI[A^+]} \right|^2, \qquad (11.26)$$

explaining the relation (11.22). An extension of this result to manifolds with boundary would require a better understanding of the relationship between the Hilbert space $\mathcal{H}_{TV}$ and the corresponding Chern–Simons Hilbert space at finite k. Some progress has been made in this direction, but the problem remains open.

This relationship to the first-order formalism may explain the appearance of positive- and negative-frequency components in the original Ponzano–Regge amplitude (11.7). The metric form of the three-dimensional Einstein action is invariant under the reflection $e^a \rightarrow -e^a$ of the triad. In the first-order formalism, on the other hand, the action changes sign under such a transformation. The first-order path integral – which is, after all, a sum over triads and spin connections – thus includes a pair of contributions, differing in the sign of the exponent, for each geometry that appears in the second-order path integral. For the path integral computations of chapter 9 to be valid, we must count both of these contributions; if we restrict triads to, say, those with positive determinant, we will not obtain the delta functionals needed in those calculations.

5. As noted earlier, it is not easy to find a discrete version of the Wheeler–DeWitt equation. Recently, however, Barrett and Crane have found a plausible candidate, and have shown that the wave functions of Ponzano and Regge satisfy the resulting equation [28].

6. The connected sum formula (9.74) for a topological field theory has been checked in the Turaev–Viro model by direct computation [157].

As in the Euclidean path integral approach of chapter 10, one can also consider the sum over topologies in the Turaev–Viro model. In analogy to the matrix models of two-dimensional gravity, Boulatov has found a generating function, a path integral of the form

$$Z = \int [d\phi] e^{-I[\phi]} \tag{11.27}$$

with fields $\phi(x, y, z)$, $x, y, z \in U_q(sl(2))$, that yields a weighted sum of Turaev–Viro partition functions over all simplicial complexes [45, 217]. The resulting sum includes complexes that are not manifolds, and it is almost certainly divergent; however, it may be possible to restrict the sum to a suitably limited class of topologies.

These results demonstrate a very close relationship between the lattice formulation of Ponzano and Regge and a variety of other quantizations of three-dimensional Euclidean gravity. The most important outstanding question is whether a Lorentzian analog can be found. The geometry of Lorentzian metrics is quite different from that of Euclidean metrics, and in the first-order formalism, the representation theory of $SO(2, 1)$ differs dramatically from that of $SU(2)$. Barrett and Foxon have taken an important first step in understanding the Lorentzian version of the Ponzano–Regge amplitude, but a great deal of work remains [29].

11.3 A Hamiltonian lattice formulation

The lattice methods of the preceding sections were manifestly covariant, requiring no splitting of spacetime into space and time. Just as in the continuum theory, however, one can also develop a Hamiltonian lattice approach, which describes the evolution of a two-dimensional spatial lattice in time. One version of this model, related to the first-order continuum formalism, has been studied by Waelbroeck and his collaborators [84, 266, 267, 268, 269, 272]. A related model, a kind of gauge-fixed Hamiltonian lattice formalism, has been developed by 't Hooft and his colleagues; this approach will be discussed in the next section.

Consider a two-dimensional spatial lattice, not necessarily planar, with faces $\{F_i\}$, vertices $\{V_I\}$, and edges $\{E_{ij}\}$, where the edge E_{ij} lies between faces F_i and F_j. We can think of this lattice as sitting in a $(2+1)$-dimensional spacetime, and can assign a Lorentzian triad $e^a(i)$ to each face F_i. To each edge E_{ij}, Waelbroeck assigns the following variables:

1. a matrix $M^a{}_b(ij) \in SO(2,1)$, which gives the Lorentz transformation relating the triads $e^b(j)$ and $e^a(i)$, and

2. a vector $E^a(ij)$, which can be interpreted as the components of the edge E_{ij}, viewed as a spacetime vector, in the frame $e^a(i)$.

These variables are the lattice analogs of the loop variables T^0 and T^1 of equations (4.81)–(4.82), or more precisely the untraced versions of these variables. In particular, $M(ij)$ is essentially the (untraced) Wilson line between F_i and F_j in the spin 1 representation.

Waelbroeck's variables are not all independent, but must satisfy the relations

$$M^a{}_c(ij)M^c{}_b(ji) = \delta^a_b$$
$$E^a(ji) = -M^a{}_b(ij)E^b(ij). \qquad (11.28)$$

Poisson brackets for these quantities can be guessed from the corresponding brackets (4.84) for the loop variables, or deduced from the hypothesis that the infinitesimal generators of the $M(ij)$ are canonically conjugate to the $E(ij)$. One finds

$$\left\{ E^a(ij), E^b(ij) \right\} = \epsilon^{abc} E_c(ij)$$
$$\left\{ E^a(ij), M^b{}_c(ij) \right\} = \epsilon^{ab}{}_d M^d{}_c(ij)$$
$$\left\{ E^a(ij), M^b{}_c(ji) \right\} = -\epsilon^{ad}{}_c M^b{}_d(ij), \qquad (11.29)$$

with all other brackets vanishing.

We must next implement the constraints (2.98) of the first-order formalism. The torsion constraint $\mathscr{C}^a = 0$ can be interpreted as the requirement that each face of the lattice close:

$$J^a(i) = \sum_{E_{ij} \in \partial F_i} E^a(ij) = 0. \tag{11.30}$$

The curvature constraint $\tilde{\mathscr{C}}^a = 0$ is the requirement that the product of Lorentz transformations around each vertex be the identity, that is, that there be no deficit angle at any vertex:

$$W^{a_1}{}_{a_{n+1}}(I) = M^{a_1}{}_{a_2}(i_1 i_2) M^{a_2}{}_{a_3}(i_2 i_3) \dots M^{a_n}{}_{a_{n+1}}(i_n i_1) - \delta^{a_1}_{a_{n+1}} = 0, \tag{11.31}$$

where $F_1, \dots, F_n$ are the faces surrounding vertex V_I. Using the brackets (11.29), it may be shown that these constraints generate the Lorentz transformations and local translations of the lattice variables $E^a(ij)$ and $M^a{}_b(ij)$, and that they satisfy the discrete analog of the $ISO(2,1)$ commutation relations (2.102). Moreover, Waelbroeck has shown that the continuum limit of the action for this lattice theory is precisely the first-order action (2.61) for (2+1)-dimensional gravity with a vanishing cosmological constant.

It is a simple exercise to count the degrees of freedom in this model. There are six phase space degrees of freedom for each edge, three components of $E^a(ij)$ and three of $M^a{}_b(ij)$; three constraints J^a for each face; and three constraints $W^a{}_b$ for each vertex. The constraints are first class, and must be counted twice: each constraint eliminates one variable directly, but also generates a gauge transformation that can be used to eliminate another. The number of remaining physical degrees of freedom is thus

$$N = 6(\textit{number of edges}) - 6(\textit{number of faces}) - 6(\textit{number of vertices})$$
$$= -6\chi = 6(2g - 2), \tag{11.32}$$

where χ is the Euler number and g is the genus of the surface formed by the lattice. As we saw in chapter 2, N is the correct number of degrees of freedom needed to parametrize the reduced phase space for (2+1)-dimensional gravity on a spacetime $[0, 1] \times \Sigma_g$. The case of the torus is a bit trickier, since the lattice constraints are not all independent, but the correct number of degrees of freedom can again be obtained. Point particles – vertices at which a deficit angle is present and the constraint (11.31) is thus violated – can also be incorporated into the model, and the counting of degrees of freedom is again correct.

Observe that the number N is independent of the form of the lattice, and can be determined from the overall spatial topology. This is a reflection of extensive gauge freedom available in the choice of lattice. In particular, one can choose a gauge in which all but a small number of the lattice vectors $E^a(ij)$ are zero, reducing the effective lattice size.

The simplest gauge choice[§] is one that reduces the lattice to the $4g$-sided polygon $P(\Sigma)$ described in chapter 4 and appendix A (see figure 4.1). The remaining independent variables are $E^a(\mu)$ and $M^a{}_b(\mu)$, $\mu = 1, 2, \ldots, 2g$, satisfying the constraints

$$\sum_\mu \left[E^a(\mu) - E^b(\mu) M_b{}^a(\mu) \right] = 0, \tag{11.33}$$

$$\epsilon^{abc} \left[M(1)M(2)M(1)^{-1}M(2)^{-1} \ldots \right.$$
$$\left. M(2g-1)M(2g)M(2g-1)^{-1}M(2g)^{-1} \right]_{bc} = 0.$$

These variables obey simple equations of motion, and the classical solutions of the vacuum field equations are easy to find: one can choose a coordinate system in which

$$M^a{}_b(\mu)(t) = M^a{}_b(\mu)(0)$$
$$E^a(\mu)(t) = E^a(\mu)(0) + tV^a(\mu)(0), \tag{11.34}$$

with constants subject to the constraints (11.33). The resulting picture is exactly that of chapter 4: the spacetime can be visualized as a polygonal tube cut out of Minkowski space, with corners lying on straight world lines and edges identified pairwise.

The lattice model developed here is closely related to the geometric structure approach described in chapter 4, and it may be quantized according to the prescription of chapter 6. For the torus, for example, the relationship between the $E^a(\mu)$ and $M^a{}_b(\mu)$ and the holonomy variables $\{\lambda, \mu, a, b\}$ has been worked out explicitly [272]. Note that the $E^a(\mu)$ and $M^a{}_b(\mu)$ are not quite gauge invariant, but change nontrivially under the transformations generated by the constraints (11.33). In the covariant canonical quantization of chapter 6, these variables would become families of 'perennials' analogous to the operators $\hat{\tau}(T)$ of that chapter.

Alternatively, we can approach the remaining constraints (11.33) in the way we treated the constraints in the ADM formalism: we can choose some particular combination of the $E^a(\mu)$ to use as a time variable, solve the constraints, and work on the corresponding reduced phase space. Waelbroeck has suggested the choice

$$T = -\frac{E^1(1)}{M^2{}_0(1)}, \tag{11.35}$$

[§] Waelbroeck and Zapata have also discussed the possibility of choosing a triangular lattice and relating the resulting model to Regge calculus [270].

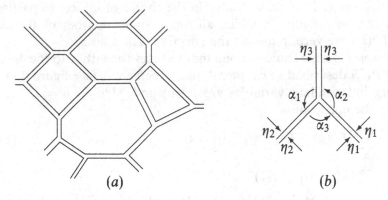

Fig. 11.5.   In 't Hooft's approach to (2+1)-dimensional gravity, a Cauchy surface is decomposed into a collection of flat polygons. Figure (b) illustrates the rapidities η and the angles α near a vertex.

for example, and has found some results for the corresponding quantum theory [269].   Observe that by (11.34), T is proportional to a simple classical time coordinate.   A slightly different choice of T is needed for the torus, as described in reference [84].   Waelbroeck has argued that mapping class group invariance can be imposed by summing over images, but the questions of convergence of the resulting series and of the appropriate inner product have not yet been worked out, and the relationship between this approach and other methods of quantization remains poorly understood.

11.4 't Hooft's polygon model

The lattice model of the preceding section was based on the first-order form of the Einstein action. A different Hamiltonian lattice model, based on the metric formalism, has been developed over the past few years by 't Hooft and others [112, 248, 249, 250, 251, 252, 280, 281]. Classically, 't Hooft's polygon model can be viewed as a gauge-fixed version of the lattice formulation of Waelbroeck *et al.*, but the two theories are related by a nonlocal change of variables, so it is not clear that the quantum theories should be equivalent.

't Hooft starts with a Cauchy surface Σ_t labeled by a time coordinate t, and decomposes it into a collection of flat polygons, chosen so that at most three edges meet at each vertex, as illustrated in figure 11.5. Locations of point particles are considered to be vertices connected to a single edge. If Σ_t is a surface of genus $g \neq 1$, it does not admit a globally flat metric, of course, but a piecewise flat metric always exists; as in Regge calculus, the spatial curvature is then concentrated entirely at the vertices. As in the

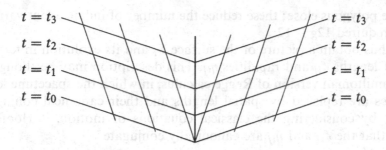

Fig. 11.6. The Lorentz frames associated with two adjacent polygons are different, and, unintuitively, a single edge – shown here as a vertical line – may move in opposite directions with respect to two neighboring frames.

preceding section, each polygon carries an associated frame, but we now assume that the coordinate t is the time coordinate in every frame, thus partially fixing the gauge.

In a static spacetime, the edges of each polygon remain fixed. In a general spacetime, however, an edge may move at a constant velocity, and its length may change. In the rest frame of a polygon F, the motion of an edge may be represented by a Lorentz transformation with boost parameter (or rapidity) η, or velocity $\tanh\eta$. A polygonal decomposition is thus labeled by a set of edge lengths, angles, and rapidities, as in figure 11.5. It is important to keep in mind that a figure of this sort is an abbreviation that does not show the full geometry. For example, it is possible – and, indeed, standard – for an edge E_{ij} between adjacent faces F_i and F_j to move inward (or outward) with respect the frames of both polygons, as shown in figure 11.6.

The lengths, angles, and boosts of figure 11.5 are not all independent: they are restricted by the requirements of local flatness and by the choice of time coordinate t. In particular, it may be shown that the length ℓ_{ij} of an edge E_{ij} must be the same when measured in either of the two frames associated with the adjacent polygons F_i and F_j; the velocity of any edge must be orthogonal to that edge; and the boost parameters η_{ij} and η_{ji} of any pair of identified edges must be equal in magnitude (but typically opposite in direction, as in figure 11.6).

An additional set of 'triangle conditions' comes from the requirement that in the absence of matter, the full (three-dimensional) curvature vanish at each vertex. This implies that the product of Lorentz transformations around any such vertex must be the identity. These conditions are equivalent to the constraints (11.31) in Waelbroeck's lattice model, and it may be shown that they determine the angles α_i of figure 11.5 in terms of the rapidities η_{ij}. Additional constraints occur in the form of the requirement that the polygons close; these reduce the number of independent variables to the required $12g - 12$.

We thus obtain a picture of the surface Σ_t and its evolution in terms of a set of lengths ℓ_{ij} and rapidities η_{ij}. This description may be thought of as a Hamiltonian version of Regge calculus, in which the spacetime length variables are replaced by spatial lengths and their canonical conjugates. Indeed, by considering the classical equations of motion, 't Hooft has shown that the ℓ_{ij} and η_{ij} are canonically conjugate,

$$\{2\eta_{ij}, \ell_{kl}\} = \delta_{(ij),(kl)}. \tag{11.36}$$

The corresponding Hamiltonian is the sum of deficit angles around the vertices,

$$H = \sum_{\text{vertices}:I} \left(2\pi - \sum_{F_i \text{ meets } V_I} \alpha_i \right). \tag{11.37}$$

The angles α_i are determined by the rapidities η_{ij}, and their Poisson brackets may be computed from (11.36); the resulting equations of motion,

$$\frac{d\ell_{ij}}{dt} = \{H, \ell_{ij}\}, \qquad \frac{d\eta_{ij}}{dt} = \{H, \eta_{ij}\} = 0, \tag{11.38}$$

agree with the classical vacuum field equations of (2+1)-dimensional gravity. The polygonal lattice formulation of this section may obtained from the lattice model of the preceding section by a partial gauge-fixing and a nonlocal redefinition of variables [271]. It is also clearly similar to the description of geometric structures developed in chapter 4, although the details of this relationship have not been worked out.

A given polygonal decomposition such as that of figure 11.5 is not stable under time evolution: an edge length may shrink to zero, and a vertex may collide with an edge. This will lead to changes in the topological structure of the lattice, such as those pictured symbolically in figure 11.7. The possible changes in the lattice structure have been enumerated in reference [250] – there are nine topologically distinct possibilities, five if no point sources are present – and the corresponding changes in lengths and rapidities have been completely worked out. The resulting evolution may be simulated on a computer, providing a powerful method for visualizing classical evolution in 2+1 dimensions. In particular, computer simulations are potentially a valuable tool for understanding the structure of the initial or final singularity for a universe with the spatial topology of a genus $g > 1$ surface.

To quantize this model, it is natural to start with the torus topology $[0, 1] \times T^2$, as we have in previous chapters. Some spacetimes with

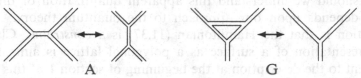

Fig. 11.7. There are nine possible changes in the lattice topology in 't Hooft's lattice model. Two of them (called *A* and *G* by 't Hooft) are illustrated here.

this topology can be described by a single polygon with a simple set of variables. Unfortunately, however, it appears that two polygons are necessary to parametrize the full moduli space, and the relationships among the length and rapidity variables become quite complicated [280]. Further complications arise because the variables ℓ_{ij} and η_{ij} obey a set of inequalities – positivity of length and the triangle inequality. Such inequalities are difficult to implement in a quantum theory. For example, one cannot represent the Poisson brackets (11.36) as standard canonical commutators of self-adjoint operators, $[2\hat{\eta}_{ij}, \hat{\ell}_{kl}] = i\delta_{(ij),(kl)}$: in such a representation, the unitary operator

$$U_\alpha = \exp\{i\alpha\hat{\eta}_{ij}\} \tag{11.39}$$

could generate translations of ℓ_{ij} to negative values [158].

Nevertheless, even in the absence of a fully worked-out solution one can make some interesting observations concerning the quantization of the polygonal lattice model. In particular, 't Hooft has stressed that the Hamiltonian (11.37) is a linear combination of angles, and as such is multivalued, with H indistinguishable from $H + 2\pi$. This observation is reinforced by an analysis of the dynamics: the Hamiltonian only appears in the combinations $\cos H$ and $\sin H$.

In ordinary quantum mechanics, the observable conjugate to such a multivalued variable is typically discrete. Here, the behavior of H suggests that time should be discrete: if H is multivalued, the time translation operator

$$U_t = \exp\{it\hat{H}\} \tag{11.40}$$

is well defined only when t is an integer. It thus seems that time should be quantized in Planck units. Such a discretization does not seem to occur for spatial coordinates. For the case of a single point particle, however, 't Hooft has found a new set of variables – essentially the coordinates of the particle with respect to a fixed origin – whose conjugate momenta are also angles, leading to a possible quantization of space as well as time. The resulting $S^1 \times S^2$ lattice, which admits an action of the full Lorentz group, is described in reference [252].

How should we understand this apparent quantization of time? The answer depends upon our approach to the quantum theory. The key observation is that the Hamiltonian (11.37) is a *constraint*. Classically, the representation of a surface as a polygonal lattice is almost exactly equivalent to the description at the beginning of section 1 of this chapter: each vertex V_I with deficit angle δ_I contributes an amount $2\delta_I$ to the action, and the total curvature is

$$\int d^2x \sqrt{{}^{(2)}g}\,{}^{(2)}R = 2\sum \delta_I = 2H. \qquad (11.41)$$

But for a closed surface Σ, the Gauss–Bonnet theorem tells us that

$$\int d^2x \sqrt{{}^{(2)}g}\,{}^{(2)}R = 2\pi\chi(\Sigma), \qquad (11.42)$$

where $\chi = 2 - 2g$ is the Euler number. H is therefore proportional to the genus of Σ,

$$H = 2\pi(1 - g). \qquad (11.43)$$

Thus as long as we consider the global topology of spacetime to be observable, H is not really multivalued, even though it has been expressed as a sum of angles; after all, a shift $g \to g + 1$ should be observable.

If the quantization of the polygonal lattice model is viewed as a gauge-fixed version of the lattice quantum theory of section 3, then equation (11.43) continues to hold as a first class constraint. The gauge transformations generated by this constraint are a part of the symmetry used to fix the time coordinate t. Now, in the standard approach to quantizing a constrained system, the reduced phase space is obtained by *both* fixing the gauge symmetry *and* solving the corresponding constraints, thus fixing the value of H. As Waelbroeck and Zapata have shown, one can then go on to define an 'internal time', analogous to the York time TrK of chapter 5; the corresponding physical Hamiltonian, analogous to H_{red}, is single valued, and there is no reason to suppose time is discrete [271].

The situation changes if we consider an open universe. If Σ is an open surface, the Hamiltonian (11.37) may be interpreted as a boundary Hamiltonian of the type discussed in sections 4 and 7 of chapter 2. For a single point source, for instance, the generator $\bar{G}$ of time translations discussed in chapter 3 is precisely the Hamiltonian (11.37) for a surface with a single vertex. Such a boundary Hamiltonian is not constrained to vanish, and it generates translations in an observable time parameter, the time measured by an observer at rest at infinity.

In this setting, the argument for discrete time is more plausible. There remains a choice, however, of whether H and $H + 2\pi$ should be considered distinguishable or indistinguishable. This choice is related to an ambiguity in the gauge group in the first order formalism: the group may be either $ISO(2,1)$, for which a periodic direction occurs, or its universal cover $\widetilde{ISO}(2,1)$. In some approaches to the quantization of point particles there is evidence that the group $\widetilde{ISO}(2,1)$ is required to obtain the correct classical limit, but this conclusion may not hold in other approaches to the quantum theory [51].

We might hope to learn more about this issue by studying (2+1)-dimensional quantum gravity coupled to matter fields. Fairly little is known about such systems, but if we restrict our attention to spacetimes with the topology $\mathbf{R}^3$ and to circularly symmetric metrics, a well-defined 'midi-superspace' quantization exists for gravity coupled to a scalar field [16]. Given a reasonable (although not unique) choice of operator ordering, Ashtekar and Pierri have shown that the Hamiltonian analogous to (11.37) for this system has a spectrum $[0, 2\pi]$. In particular, although the Hamiltonian has a classical interpretation as a deficit angle, it has no eigenvalues greater than 2π, and is not multivalued. Once again, however, the quantum theory upon which this analysis is based is not unique – there are other choices of operator ordering for which the spectrum of the Hamiltonian is quite different.

11.5 Dynamical triangulation

The lattice models we have seen so far are based on a fixed triangulation of space or spacetime, with edge lengths serving as the basic gravitational variables. An alternative scheme is the 'dynamical triangulation' model, in which edge lengths are fixed and the path integral is represented as a sum over triangulations.[¶] This approach has been proven to be quite useful in two-dimensional gravity, and some progress has been made in the higher-dimensional analogs.

The starting point for the dynamical triangulation model is a simplicial complex, diffeomorphic to a manifold M, composed of an arbitrary number of equilateral tetrahedra with sides of length a. Metric information is no longer contained in the choice of edge lengths, but rather depends on the combinatorial pattern of the tetrahedra. Unlike the approaches described above, the dynamical triangulation model is not exact in 2+1 dimensions, but one might hope that as a becomes small and the number of tetrahedra becomes large it may be possible to approximate an arbitrary geometry.

Let N_0, N_1, N_2, and N_3 be the number of vertices, edges, faces, and tetrahedra in a given triangulation. These are not all independent: one

---

¶ For a review of this approach in two, three, and four dimensions, see reference [5].

has

$$\chi(M) = N_0 - N_1 + N_2 - N_3 = 0,$$
$$N_2 = 2N_3, \tag{11.44}$$

where the first equality is the statement that the Euler number of any odd-dimensional manifold vanishes and the second follows from the fact that each face is shared by exactly two tetrahedra. Each tetrahedron now has the same volume, yielding a total volume of

$$\int d^3x \sqrt{g} = \frac{a^3}{6\sqrt{2}} N_3. \tag{11.45}$$

The integral of the scalar curvature can be computed as in equation (11.3), yielding

$$\int d^3x \sqrt{g}\, R = a \left[2\pi N_1 - 6N_3 \cos^{-1} \left(\frac{1}{3} \right) \right]. \tag{11.46}$$

The total Einstein action, with a cosmological constant, thus takes the form

$$I_{dyn.\ triang.} = \beta N_3 - \kappa N_1 = \mu N_3 - \alpha N_0, \tag{11.47}$$

where the constants (β, κ) or (μ, α) are related to G and Λ.

The partition function may thus be written in the form

$$Z_M = \sum_{\mathcal{T}} e^{\kappa N_1 - \beta N_3}, \tag{11.48}$$

where $\mathcal{T}$ is the set of simplicial manifolds with the given topology M. As in Regge calculus, the appropriate weighting factor in this sum – the discrete analog of the path integral measure, with the right Faddeev–Popov determinants – is not clear, but one might hope for some kind of universal behavior that would make this ambiguity unimportant. In order for this sum to converge, the number of triangulations of M must not grow too fast. Indeed, we can rewrite (11.48) as

$$Z_M = \sum_{N_1, N_3} Z_{N_1 N_3}(M) e^{\kappa N_1 - \beta N_3}, \tag{11.49}$$

where $Z_{N_1 N_3}(M)$ is the number of ways M may be triangulated with N_1 edges and N_3 tetrahedra. Unless $Z_{N_1 N_3}(M)$ is exponentially bounded, this sum will clearly diverge. The existence of such an 'entropy bound' has been the subject of some debate, but there is good numerical evidence for a bound, and there now appears to be a proof [49, 50]. Interestingly, the bound on $Z_{N_1 N_3}(M)$ found by Carfora and Marzuoli involves the

Ray–Singer torsion, suggesting a connection with the path integrals of chapters 9 and 10, but this relationship has not yet been elaborated.

The sum (11.48) may be evaluated numerically, using Monte Carlo techniques [73]. These methods require a procedure for generating a sample sequence of simplicial complexes through local updates or 'moves' that change the triangulation. Several choices of moves have been shown to be ergodic in the space of triangulations of M: that is, any two configurations can be connected by a finite number of moves, as is required for the numerical methods to work [138].

A number of numerical results exist in the literature (see reference [5] and references therein; see also [139] for some numerical results in Regge calculus.) These computations indicate that as the parameters of the action (11.47) are varied, a phase transition occurs between a 'crumpled' phase, in which the universe is very small and has a very large Hausdorff dimension, and a 'branched polymer' phase, in which the Hausdorff dimension is close to two. The hope is that the physically relevant region occurs at the location of the phase transition. In contrast to the well understood two-dimensional random triangulation model, however, the continuum limit of the three-dimensional model is poorly understood, and the relationship between the numerical simulations and other approaches to quantization remains an open problem.

12
The (2+1)-dimensional black hole

The focus of the past few chapters has been on three-dimensional quantum cosmology, the quantum mechanics of spatially closed (2+1)-dimensional universes. Such cosmologies, although certainly physically unrealistic, have served us well as models with which to explore some of the ramifications of quantum gravity. But there is another (2+1)-dimensional setting that is equally useful for trying out ideas about quantum gravity: the (2+1)-dimensional black hole of Bañados, Teitelboim, and Zanelli [23] introduced in chapter 2. As we saw in that chapter, the BTZ black hole is remarkably similar in its qualitative features to the realistic Schwarzschild and Kerr black holes: it contains genuine inner and outer horizons, is characterized uniquely by an ADM-like mass and angular momentum, and has a Penrose diagram (figure 3.2) very similar to that of a Kerr–anti-de Sitter black hole in 3+1 dimensions.

In the few years since the discovery of this metric, a great deal has been learned about its properties. We now have a number of exact solutions describing black hole formation from the collapse of matter or radiation, and we know that this collapse exhibits some of the critical behavior previously discovered numerically in 3+1 dimensions. We understand a good deal about the interiors of rotating BTZ black holes, which exhibit the phenomenon of 'mass inflation' known from 3+1 dimensions. Black holes in 2+1 dimensions can carry electric or magnetic charge, and can be found in theories of dilaton gravity. Exact multi-black hole solutions have also been discovered.

In this chapter, we shall concentrate on the quantum mechanical and thermodynamic properties of the BTZ black hole. We shall investigate the (2+1)-dimensional analog of Hawking radiation, explore black hole thermodynamics, and examine a possible microscopic explanation for black hole entropy. For a broader review of classical and quantum properties of the BTZ black hole and a fairly large bibliography, the reader may wish to consult reference [62].

Following the conventions of reference [23], this chapter will use units $8G = 1$ unless otherwise stated.

12.1 A brief introduction to black hole thermodynamics

According to the classical 'no hair' theorems of general relativity, the properties of a black hole are completely determined by its mass, angular momentum, and a few conserved charges. A classical black hole state thus appears to be very simple. But as Bekenstein pointed out in 1972, a massive star is an extremely complicated system, with a large entropy that seems to be lost in its collapse to a black hole. Similarly, if we drop a box of gas into a black hole, we end up with a final state consisting merely of a slightly larger black hole, and entropy seems again to have disappeared. If we wish such processes to be consistent with the second law of thermodynamics, we must attribute a large entropy to a black hole itself [32].

By considering a variety of thought experiments, Bekenstein concluded that this entropy should be proportional to the area of black hole horizon, and argued that the constant of proportionality ought to be of order unity in Planck units. The identification of the horizon area with entropy was strengthened by the striking analogies between the 'laws of black hole mechanics' and those of thermodynamics: for example, black hole area, like entropy, can never decrease in time. The thermodynamic properties of black holes were firmly established when Hawking showed in 1973 that black holes are black bodies, radiating at a temperature

$$T_0 = \frac{\kappa}{2\pi} \tag{12.1}$$

where κ is the surface gravity [143, 144]. This result allowed an exact determination of Bekenstein's unknown proportionality constant: the entropy of a black hole with horizon area A is

$$S = \frac{1}{4}A. \tag{12.2}$$

There are now half a dozen independent derivations of Hawking radiation and the Bekenstein–Hawking entropy, and black hole thermodynamics is well established. The thermodynamic analysis of black holes is not accompanied by any generally accepted statistical mechanical description, however: there is no established microscopic explanation of the thermal properties of black holes in terms of fundamental quantum states.* More-over, several important questions remain completely open, most notably the 'information loss paradox' [126]: what happens to the information carried by a quantum field in a pure state which collapses into a black hole and then disperses as Hawking radiation in, presumably, a mixed state? It might be hoped that (2+1)-dimensional gravity, which permits an exact quantum mechanical description of black holes, could suggest answers to some of these questions.

12.2 The Lorentzian black hole

If we wish to understand the quantum mechanics of the (2+1)-dimensional black hole, an obvious preliminary question is whether Hawking radiation occurs. This is a semiclassical question, one that depends on the proper-ties of quantum fields in a black hole background but does not require a full treatment of quantum gravity. The starting point for the quan-tum field theoretical computation is an appropriate two-point function $G(x, x') = \langle 0|\phi(x)\phi(x')|0\rangle$, from which such quantities as the expectation values $\langle 0|T_{\mu\nu}|0\rangle$ can be derived. In particular, a Green's function that is periodic in imaginary time with period β is a thermal Green's function corresponding to a local inverse temperature β, and such a periodicity can be interpreted as an indication of Hawking radiation.

The properties of the Green's function $G(x, x')$ can be computed by brute force, but it is simpler and more instructive to take advantage of a peculiar property of black holes in 2+1 dimensions, the fact that they are described by spaces of constant curvature. As we saw in chapter 2, the (2+1)-dimensional black hole can be represented by the metric

$$ds^2 = -N^2 dt^2 + r^2 \left(d\phi^2 + N^\phi dt\right)^2 + N^{-2} dr^2 \qquad (12.3)$$

with

$$N^2 = -M + \frac{r^2}{\ell^2} + \frac{J^2}{4r^2}, \qquad N^\phi = -\frac{J}{2r^2}, \qquad (12.4)$$

where M and J are the mass and angular momentum. But as a space of constant negative curvature, the BTZ black hole is also locally isometric to anti-de Sitter (adS) space. In the language of chapter 4, the black hole has an anti-de Sitter geometric structure, and should be expressible as a set of adS patches glued together by suitable $SL(2, \mathbf{R}) \times SL(2, \mathbf{R})$ holonomies. This means that the Green's function $G(x, x')$ can be described in terms of the much simpler Green's function in adS space.

---

* There has been some very recent progress in finding a string theoretical description of black hole statistical mechanics [151, 246].

To derive the geometric structure of the BTZ black hole, recall from chapter 4 that anti-de Sitter space can be obtained from flat $\mathbf{R}^{2,2}$, with coordinates (X_1, X_2, T_1, T_2) and metric

$$dS^2 = dX_1^2 + dX_2^2 - dT_1^2 - dT_2^2, \qquad (12.5)$$

by restricting to the submanifold

$$X_1^2 + X_2^2 - T_1^2 - T_2^2 = -\ell^2 \qquad (12.6)$$

with the induced metric. The relevant region of the universal covering space $\widetilde{adS}$ of anti-de Sitter space may be covered by an infinite set of Kruskal coordinate patches of three types, corresponding to the regions of the Penrose diagram of figure 3.2 [22]:

I. $(r \geq r_+)$

$$X_1 = \ell\sqrt{\alpha}\sinh\left(\frac{r_+}{\ell}\phi - \frac{r_-}{\ell^2}t\right), \quad X_2 = \ell\sqrt{\alpha-1}\cosh\left(\frac{r_+}{\ell^2}t - \frac{r_-}{\ell}\phi\right)$$

$$T_1 = \ell\sqrt{\alpha}\cosh\left(\frac{r_+}{\ell}\phi - \frac{r_-}{\ell^2}t\right), \quad T_2 = \ell\sqrt{\alpha-1}\sinh\left(\frac{r_+}{\ell^2}t - \frac{r_-}{\ell}\phi\right)$$

II. $(r_- \leq r \leq r_+)$

$$X_1 = \ell\sqrt{\alpha}\sinh\left(\frac{r_+}{\ell}\phi - \frac{r_-}{\ell^2}t\right), \quad X_2 = -\ell\sqrt{1-\alpha}\sinh\left(\frac{r_+}{\ell^2}t - \frac{r_-}{\ell}\phi\right)$$

$$T_1 = \ell\sqrt{\alpha}\cosh\left(\frac{r_+}{\ell}\phi - \frac{r_-}{\ell^2}t\right), \quad T_2 = -\ell\sqrt{1-\alpha}\cosh\left(\frac{r_+}{\ell^2}t - \frac{r_-}{\ell}\phi\right)$$

III. $(0 \leq r \leq r_-)$

$$X_1 = \ell\sqrt{-\alpha}\cosh\left(\frac{r_+}{\ell}\phi - \frac{r_-}{\ell^2}t\right), \quad X_2 = -\ell\sqrt{1-\alpha}\sinh\left(\frac{r_+}{\ell^2}t - \frac{r_-}{\ell}\phi\right)$$

$$T_1 = \ell\sqrt{-\alpha}\sinh\left(\frac{r_+}{\ell}\phi - \frac{r_-}{\ell^2}t\right), \quad T_2 = -\ell\sqrt{1-\alpha}\cosh\left(\frac{r_+}{\ell^2}t - \frac{r_-}{\ell}\phi\right)$$

$$(12.7)$$

where

$$\alpha(r) = \left(\frac{r^2 - r_-^2}{r_+^2 - r_-^2}\right), \quad \phi \in (-\infty, \infty), \ t \in (-\infty, \infty). \qquad (12.8)$$

It is straightforward to show that the standard adS metric dS^2 then transforms to the BTZ metric (12.3)–(12.4) in each patch. The 'angle' ϕ in equation (12.7) has infinite range, however; to make it into a true angular variable, we must identify ϕ with $\phi + 2\pi$. This identification is an isometry of anti-de Sitter space – it is a boost in the X_1–T_1 and X_2–T_2 planes – and as in chapter 4, it corresponds to a calculable element (ρ_L, ρ_R) of $SL(2,\mathbf{R}) \times SL(2,\mathbf{R})/\mathbf{Z}_2$. Indeed, a simple computation shows that if we set

$$
\rho_L = \begin{pmatrix} e^{\pi(r_+ - r_-)/\ell} & 0 \\ 0 & e^{-\pi(r_+ - r_-)/\ell} \end{pmatrix},
$$

$$
\rho_R = \begin{pmatrix} e^{\pi(r_+ + r_-)/\ell} & 0 \\ 0 & e^{-\pi(r_+ + r_-)/\ell} \end{pmatrix}, \tag{12.9}
$$

and define $\mathbf{X}$ as in equation (4.18), then the transformation $\mathbf{X} \to \rho_L^n \mathbf{X} \rho_R^n$ corresponds to the shift $\phi \to \phi + 2\pi n$.

The BTZ black hole is thus a quotient space $\widetilde{adS}/\langle(\rho_L, \rho_R)\rangle$, where $\langle(\rho_L, \rho_R)\rangle$ denotes the group generated by (ρ_L, ρ_R). Note that as in the cosmological solutions discussed earlier, the holonomies (ρ_L, ρ_R) may be identified with the holonomies of a suitable Chern–Simons connection A, where from equation (2.77), the relevant $SL(2,\mathbf{R}) \times SL(2,\mathbf{R})$ connection is

$$
A^{(\pm)a} = \omega^a \pm \frac{1}{\ell} e^a. \tag{12.10}
$$

The group elements (12.9) were first found, in a somewhat more complicated conjugated form, by Cangemi et al., who pointed out that the mass and angular momentum M and J have a natural interpretation in terms of the two quadratic Casimir operators of $SL(2,\mathbf{R}) \times SL(2,\mathbf{R})$ [47]. Steif has investigated a similar quotient construction for the supersymmetric (2+1)-dimensional black hole [245].

Now let $G_A(x, x')$ be a two-point function on the universal covering space $\widetilde{adS}$ of anti-de Sitter space. The corresponding two-point function for the BTZ black hole can then be expressed by the method of images as

$$
G_{BTZ}(x, x') = \sum_n e^{-i\delta n} G_A(x, \Lambda^n x'), \tag{12.11}
$$

where $\Lambda x'$ denotes the action of the group element (12.9) on x'. The phase δ is zero for ordinary ('untwisted') fields, but may in principle be arbitrary, corresponding to boundary conditions $\phi(\Lambda x) = e^{-i\delta}\phi(x)$; the choice $\delta = \pi$ leads to conventional 'twisted' fields. Our problem has thus been effectively reduced to the comparatively simple task of understanding quantum field theory on $\widetilde{adS}$.

While quantum field theory on anti-de Sitter space is fairly straightforward, it is by no means trivial. The main difficulty comes from the fact that neither anti-de Sitter space nor its universal covering space are globally hyperbolic. As is evident from the Penrose diagrams of figure 3.2, spatial infinity is timelike, and information may enter or exit from the 'boundary' at infinity. One must consequently impose boundary conditions at infinity to formulate a sensible field theory.

This problem has been analyzed carefully in 3+1 dimensions by Avis, Isham, and Storey, who show that there are three reasonable boundary conditions for a scalar field at spatial infinity: Dirichlet (D), Neumann (N), and 'transparent' (T) boundary conditions [19]. The last – essentially a linear combination of Dirichlet and Neumann conditions – are most easily obtained by viewing $\widetilde{adS}$ as half of an Einstein static universe; physically they correspond to a particular recirculation of momentum and angular momentum at spatial infinity. The same choices exist in 2+1 dimensions. In particular, for a massless, conformally coupled scalar field, described by the action

$$I = -\int d^3x \sqrt{-g} \left(\frac{1}{2} g^{\mu\nu} \partial_\mu \phi \partial_\nu \phi + \frac{1}{16} R\phi^2 \right), \qquad (12.12)$$

the adS Green's functions are

$$G_A^{(D)} = \frac{1}{4\pi\sqrt{2}} \left\{ \sigma^{-1/2} - [\sigma + 2\ell^2]^{-1/2} \right\}$$

$$G_A^{(N)} = \frac{1}{4\pi\sqrt{2}} \left\{ \sigma^{-1/2} + [\sigma + 2\ell^2]^{-1/2} \right\}$$

$$G_A^{(T)} = \frac{1}{4\pi\sqrt{2}} \, \sigma^{-1/2}. \qquad (12.13)$$

Here $\sigma(x, x')$ is the square of the distance between x and x' in the embedding space $\mathbf{R}^{2,2}$ described above; in the coordinates of equation (12.7),

$$\sigma(x, x')$$
$$= \frac{1}{2} \left[(X_1 - X_1')^2 - (T_1 - T_1')^2 + (X_2 - X_2')^2 - (T_2 - T_2')^2 \right]. \quad (12.14)$$

These expressions should be supplemented with appropriate factors of $i\epsilon$ that determine which Green's functions they represent; for the Feynman propagator, for instance, σ should be interpreted as $\sigma + i\epsilon$. A more general Green's function with 'mixed' boundary conditions,

$$G_A^{(\alpha)} = \frac{1}{4\pi\sqrt{2}} \left\{ \sigma^{-1/2} - \alpha[\sigma + 2\ell^2]^{-1/2} \right\}, \qquad (12.15)$$

may also be considered. The corresponding BTZ Green's functions are then obtained from the sum (12.11).

The crucial observation is that the functions $G_{BTZ}(x, x')$ are periodic in imaginary time, with a period β given by

$$\beta = T_0^{-1} = \frac{2\pi r_+ \ell^2}{r_+^2 - r_-^2} = \frac{2\pi}{\kappa}, \qquad (12.16)$$

where κ is the surface gravity (3.53). To see this, let us consider the simple case of the static ($r_- = 0$) black hole, and let $x = (t = 0, r, \phi)$ and $x' = (t, r, \phi)$ in the 'Schwarzschild' coordinates (12.3). Recall from the derivation above that the group element Λ of equation (12.9) corresponds to a shift $\phi \to \phi + 2\pi$. A simple computation shows that in region I

$$\sigma(x, \Lambda^n x') = \ell^2 \left[-1 + \frac{r^2}{r_+^2} \cosh \frac{2\pi n r_+}{\ell} - \frac{r^2 - r_+^2}{r_+^2} \cosh \frac{r_+}{\ell^2} t \right], \qquad (12.17)$$

with similar results in region II. It is clear by inspection that $\sigma(x, \Lambda^n x')$ is periodic under $t \to t + i\beta$, and hence that each term in the sum (12.11) has the same periodicity. Strictly speaking, one must be careful of the $i\epsilon$ factors and the location of the branch cut needed to define $\sigma^{1/2}$. This takes a bit more work, but is not too complicated; see reference [177].

The periodicity implies, in turn, that the Green's functions in a BTZ black hole background are thermal, with temperature $T_0 = \beta^{-1}$. Indeed, a typical thermal Green's function in flat spacetime takes the form

$$G_\beta(x, x') = \frac{Tr e^{-\beta H} \phi(x) \phi(x')}{Tr e^{-\beta H}}, \qquad (12.18)$$

and the Hamiltonian H is the generator of time translations, so

$$G_\beta(x, x') = \frac{Tr \left(e^{-\beta H} \phi(x) e^{\beta H} \right) e^{-\beta H} \phi(x')}{Tr e^{-\beta H}}$$

$$= \frac{Tr \phi(\tilde{x}) e^{-\beta H} \phi(x')}{Tr e^{-\beta H}} = \frac{Tr e^{-\beta H} \phi(x') \phi(\tilde{x})}{Tr e^{-\beta H}} = G_\beta(x', \tilde{x}), \quad (12.19)$$

where $\tilde{x} = (t + i\beta, r, \phi)$. A free conformally coupled scalar field in a static (2+1)-dimensional black hole background thus behaves as if it were in a thermal bath of temperature T_0 with respect to the Hamiltonian generating translations in the adS time t of equation (12.3). Note that the relationship (12.16) of T_0 and κ is precisely the same as the corresponding result (12.1) in 3+1 dimensions.

A bit of caution is needed in interpreting this temperature. There is no obvious 'natural' scale for the anti-de Sitter time t – the scale is defined by the condition that the metric (12.3) asymptotically approach the standard anti-de Sitter metric,

$$ds^2 = (-1 + \frac{r^2}{\ell^2}) dt^2 - r^2 d\phi^2 - (-1 + \frac{r^2}{\ell^2})^{-1} dr^2, \qquad (12.20)$$

but the physical significance of such an asymptotic condition is not entirely clear. Rather, the temperature T_0 should be understood as follows. For an observer at a fixed value of r and ϕ, the proper time is Nt. A Green's function is thus periodic in imaginary proper time with period N/T_0, and this will determine the local temperature measured by such an observer. The factor $g_{00}^{-1/2}$ correcting T_0 is the standard Tolman red shift factor for the temperature in a gravitational field.

When $r_- \neq 0$, the computation is slightly more complicated, but still straightforward. One finds that σ is now periodic under the transformation

$$(t, \phi) \rightarrow (t + i\beta, \phi + i\Phi), \tag{12.21}$$

where β is again given by (12.16) and

$$\Phi = \frac{2\pi r_- \ell}{r_+^2 - r_-^2}. \tag{12.22}$$

This periodicity is the sign of a thermal Green's function with an inverse temperature β and a rotational chemical potential

$$\mu = -\Phi/\beta = -\frac{r_-}{r_+ \ell} = \Omega_H, \tag{12.23}$$

where Ω_H is the angular velocity of the horizon. Indeed, a typical Green's function in the grand canonical ensemble takes the form

$$G_{\beta,\mu}(x, x') = \frac{Tr e^{-\beta(H - \mu J)} \phi(x) \phi(x')}{Tr e^{-\beta(H - \mu J)}}, \tag{12.24}$$

where J is the generator of rotations, and the same derivation that led to equation (12.19) now yields a periodicity under $(t, \phi) \rightarrow (t + i\beta, \phi - i\beta\mu)$. Alternatively, if one defines a new angular coordinate

$$\tilde{\phi} = \phi - \frac{r_-}{\ell r_+} t, \tag{12.25}$$

it is straightforward to show that the Green's functions are again periodic under $t \rightarrow t + i\beta$. The coordinate $\tilde{\phi}$ rotates with the horizon, and it is not hard to see that the shift function $N^{\tilde{\phi}}$ vanishes at $r = r_+$.

By studying the analyticity properties of the Green's functions (12.11), Lifschytz and Ortiz argue that the relevant vacuum state is the Hartle–Hawking vacuum [177]. We may obtain further evidence for Hawking radiation by computing the Bogoliubov coefficients that transform between anti-de Sitter 'in' states in the far past and BTZ 'out' states in the far future. The relevant computations have been performed for the rotating black hole with transparent boundary conditions by Hyun, Lee, and Yee

[154], who show that the expectation value of the 'out' number operator in the 'in' vacuum state is

$$_{\text{in}}\langle 0|N_{\omega m}^{\text{out}}|0\rangle_{\text{in}} = \frac{1}{\exp\left[2\pi(\omega - m\Omega_H)\beta\right] - 1}. \tag{12.26}$$

The thermal form of equation (12.26) is again an indication of Hawking radiation; the dependence on $\omega - m\Omega_H$ rather than ω is a sign that the rotating BTZ black hole, like the Kerr black hole, has super-radiant modes. Further computations of thermodynamic behavior have been performed for a variety of field theories in black hole backgrounds [155, 156, 184, 240, 241, 244].

The semiclassical computations of this section do not directly yield a value for the entropy of the (2+1)-dimensional black hole. However, an expression for the entropy can be obtained by integrating the first law of thermodynamics,

$$dM = TdS - \mu dJ. \tag{12.27}$$

Inserting the expressions (3.46), (12.16), and (12.23) for M, J, T, and μ, we find that

$$dS = 4\pi dr_+, \tag{12.28}$$

or restoring factors of G,

$$S = \frac{2\pi r_+}{4G}. \tag{12.29}$$

Once again the results agree with the (3+1)-dimensional expression (12.2) for the Bekenstein–Hawking entropy: the entropy is one-fourth of the horizon size in Planck units.

12.3 The Euclidean black hole

The thermodynamic properties of black holes can be derived, as we have just seen, by considering the behavior of quantum fields in a classical black hole background. But as Gibbons and Hawking first pointed out, using techniques later refined by Brown and York, these properties can also be obtained from a suitable approximation of quantum gravity itself [44, 125].

In the absence of a complete quantum theory of gravity, of course, one cannot hope to compute the thermodynamic behavior of a black hole exactly. But even without a complete theory it is possible to write down a formal path integral for the gravitational partition function, which can then be approximated by the method of steepest descents described in chapter 10. In 3+1 dimensions, the leading saddle point may be shown to be the Euclidean black hole, which exists only when the Euclidean time τ has an appropriate periodicity. This periodicity, in turn, corresponds to an inverse temperature identical to that derived from quantum field theoretical calculations in a black hole background. Moreover, the saddle point contribution to the partition function gives the correct Bekenstein–Hawking entropy. In this section, we shall see that a similar computation is possible in 2+1 dimensions, and that we can even compute the one-loop correction to the entropy.

The starting point for the Gibbons–Hawking approach to black hole thermodynamics is the path integral

$$Z[\beta] = Tr e^{-\beta H} = \int [dg] e^{-I_E(\beta)}, \tag{12.30}$$

where $I_E(\beta)$ is the Euclidean gravitational action for a spacetime periodic in (imaginary) time with period β. As usual in the path integral formulation of thermodynamics, the quantity $Z[\beta]$ is to be interpreted as the canonical partition function at inverse temperature β. Brown and York have argued that the canonical formalism is not, strictly speaking, a consistent approach – black holes have negative heat capacities, making the partition function ill-defined – and that one should instead start with a microcanonical path integral for a system in a box with a finite radius [44]. For the (2+1)-dimensional black hole, however, the two methods ultimately lead to identical expressions for temperature and entropy, so for simplicity I will concentrate on the original Gibbons–Hawking approach. (See reference [83] for the application of the Brown–York formalism in 2+1 dimensions.)

Let us evaluate the path integral (12.30) in the saddle point approximation, in the topological sector that includes the Euclidean version of the BTZ black hole. The relevant metric may be obtained from (12.3)–(12.4) by letting $t = i\tau$ and $J = iJ_E$, yielding

$$ds_E^2 = N_E^2 d\tau^2 + f_E^{-2} dr^2 + r^2 \left(d\phi + N_E^\phi d\tau \right)^2 \tag{12.31}$$

with

$$N_E = f_E = \left(-M + \frac{r^2}{\ell^2} - \frac{J_E^2}{4r^2} \right)^{1/2}, \qquad N_E^\phi = -iN^\phi = -\frac{J_E}{2r^2}. \tag{12.32}$$

Note that the continuation of J to imaginary values, necessary for the metric (12.31) to be real, is physically sensible, since the angular velocity is now a rate of change of a real angle with respect to imaginary time.

The Euclidean lapse function now has roots

$$r_+ = \left\{ \frac{M\ell^2}{2} \left[1 + \left(1 + \frac{J_E^2}{M^2\ell^2} \right)^{1/2} \right] \right\}^{1/2},$$

$$r_- = -i|r_-| = \left\{ \frac{M\ell^2}{2} \left[1 - \left(1 + \frac{J_E^2}{M^2\ell^2} \right)^{1/2} \right] \right\}^{1/2}.$$

$$(12.33)$$

The metric (12.31) is a Euclidean (positive definite) metric of constant negative curvature, and the spacetime is therefore locally isometric to hyperbolic three-space $\mathbf{H}^3$. This isometry is most easily exhibited by means of the Euclidean analog of the coordinate transformation (3.59), which is now globally valid [71],

$$x = \left(\frac{r^2 - r_+^2}{r^2 - r_-^2} \right)^{1/2} \cos\left(\frac{r_+}{\ell^2}\tau + \frac{|r_-|}{\ell}\phi \right) \exp\left\{ \frac{r_+}{\ell}\phi - \frac{|r_-|}{\ell^2}\tau \right\}$$

$$y = \left(\frac{r^2 - r_+^2}{r^2 - r_-^2} \right)^{1/2} \sin\left(\frac{r_+}{\ell^2}\tau + \frac{|r_-|}{\ell}\phi \right) \exp\left\{ \frac{r_+}{\ell}\phi - \frac{|r_-|}{\ell^2}\tau \right\}$$

$$z = \left(\frac{r_+^2 - r_-^2}{r^2 - r_-^2} \right)^{1/2} \exp\left\{ \frac{r_+}{\ell}\phi - \frac{|r_-|}{\ell^2}\tau \right\}.$$

$$(12.34)$$

In these coordinates, the metric becomes that of the standard upper half-space representation of hyperbolic three-space,

$$ds_E^2 = \frac{\ell^2}{z^2}(dx^2 + dy^2 + dz^2)$$

$$= \frac{\ell^2}{\sin^2\chi}\left(\frac{dR^2}{R^2} + d\chi^2 + \cos^2\chi \, d\theta^2 \right) \qquad (z > 0),$$

$$(12.35)$$

where (R, θ, χ) are spherical coordinates for the upper half-space $z > 0$,

$$(x, y, z) = (R\cos\theta\cos\chi, R\sin\theta\cos\chi, R\sin\chi). \qquad (12.36)$$

Just as periodicity of the Lorentzian solution in the Schwarzschild angular coordinate ϕ led to the identifications (3.61), periodicity of the Euclidean solution requires that we identify

$$(R, \theta, \chi) \sim (Re^{2\pi r_+/\ell}, \theta + \frac{2\pi|r_-|}{\ell}, \chi). \qquad (12.37)$$

Equation (12.37) is the Euclidean version of the identifications $\langle(\rho_R, \rho_L)\rangle$ of section 2. A fundamental region for these identifications is the space

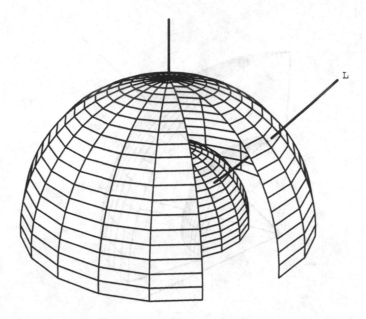

Fig. 12.1. In the upper half-space representation of $\mathbf{H}^3$, a fundamental region for the Euclidean black hole is the region between two hemispheres, with points on the inner and outer surfaces identified along lines such as L. (The outer hemisphere has been cut open here to show the inner hemisphere.)

between two hemispheres $R = 1$ and $R = e^{2\pi r_+/\ell}$, with the inside and outside boundaries identified by a translation along a radial line followed by a $2\pi|r_-|/\ell$ rotation around the z axis (see figure 12.1). Topologically, the resulting manifold is a solid torus; for $\chi \neq \pi/2$, each slice of fixed χ is an ordinary two-torus, with circumferences parametrized by the periodic coordinates $\ln R$ and θ, while the degenerate surface $\chi = \pi/2$ is a circle at the core of the solid torus. Physically, R is an angular coordinate, related to the Schwarzschild coordinate ϕ; the azimuthal angle θ measures time; and χ is a radial coordinate.

Now, for the coordinate transformation (12.34) to be nonsingular at the z axis $r = r_+$, we must require periodicity in the arguments of the trigonometric functions. That is, we must identify

$$(\phi, \tau) \sim (\phi + \Phi, \tau + \beta) \tag{12.38}$$

where

$$\beta = \frac{2\pi r_+ \ell^2}{r_+^2 - r_-^2}, \qquad \Phi = \frac{2\pi|r_-|\ell}{r_+^2 - r_-^2}. \tag{12.39}$$

If this requirement of periodicity is lifted, the Euclidean black hole acquires

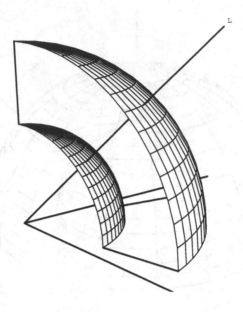

Fig. 12.2.   In the upper half-space representation of figure 12.1, Euclidean time is an angular coordinate around the z axis. This figure shows the region between two slices of constant τ.

a conical singularity at the horizon, and is no longer an exact solution of the classical Einstein field equations. Note that the periodicity (12.38)–(12.39) is identical to the periodicity (12.21) of the Lorentzian Green's functions, thus confirming the results of the preceding section.

To carry our computation of the partition function further, we need the value of the Euclidean action evaluated at the classical black hole geometry. As we saw in a slightly different context in chapter 2, this action is not merely the standard Einstein–Hilbert action, but may also require the addition of boundary terms at infinity and (for the black hole) at the horizon. In particular, consider the term

$$I_0 = \frac{1}{2\pi} \int d\tau \int_\Sigma d^2x \, N \sqrt{{}^{(2)}g} \, {}^{(2)}R \qquad (12.40)$$

in the ADM action (2.12) on a slice of constant τ, as shown in figure 12.2. (Remember that our units are now $8G = 1$.) Let us concentrate for the moment on the inner boundary $r = r_+$. The lapse function N vanishes at this boundary, of course – this is essentially the definition of the horizon – but its normal derivatives may be nonzero. A simple computation shows that the variation of the action (12.40) takes the form

$$\delta I_0 = \cdots + \frac{1}{2\pi} \int d\tau \int_{\partial \Sigma} d\phi \sqrt{\sigma} \Big[N n^k g^{ij} (\nabla_j \delta g_{ik} - \nabla_k \delta g_{ij})$$

$$- \nabla^k N n^i \delta g_{ik} + n^k \nabla_k N g^{ij} \delta g_{ij} \Big]$$

$$= \cdots + \frac{1}{2\pi} \int d\tau \int_{\partial \Sigma} d\phi \, 2(n^i \partial_i N) \delta \sqrt{\sigma}, \tag{12.41}$$

where σ is the induced metric on the boundary $\partial \Sigma$ and n^i is the radial normal to $\partial \Sigma$ in Σ. The omitted terms are the conventional 'bulk' terms that would be present even with no boundary, and in the second line I have also dropped terms that vanish because $N(r_+) = 0$.

Let us analyze this expression in more detail. Note first that the integral

$$\int n^i \partial_i N d\tau$$

must equal 2π if the geometry is smooth at the horizon. Indeed, consider an infinitesimal circle $r = r_+ + \delta r$ around the horizon. Such a circle will have a proper circumference

$$s = \int N d\tau \approx \int \delta r \partial_r N(r_+) d\tau \tag{12.42}$$

and a proper radius

$$\rho = \int_{r_+}^{r_+ + \delta r} \sqrt{g_{rr}} dr. \tag{12.43}$$

Thus

$$\frac{ds}{d\rho} = 2\pi - \Theta = \frac{1}{\sqrt{g_{rr}}} \int \partial_r N d\tau = \int n^i \partial_i N d\tau, \tag{12.44}$$

where Θ is the deficit angle at the origin. Next, observe that the integral

$$\int d\phi \sqrt{\sigma} = 2\pi r_+$$

is simply the length of the event horizon. The variation (12.41) is thus of the form

$$\delta I_0 = \cdots + 2(2\pi - \Theta) \delta r_+. \tag{12.45}$$

Now recall from section 1 of chapter 6 that the boundary terms in the variation of an action determine the symplectic form on the covariant phase space. Equation (12.45) is thus the statement that r_+ and Θ are conjugate variables. A similar conclusion can be shown to follow from the quantization of the BTZ black hole in the connection representation [71].

Suppose we now restrict our variations to those for which the geometry remains smooth at the horizon. This means that we must hold the deficit angle Θ fixed at r_+. The variational formula (12.41) tells us that to do so, we must add a term

$$ I' = -\frac{1}{2\pi} \int d\tau \int_{\partial\Sigma} d\phi\, 2(n^i \partial_i N)\sqrt{\sigma} = -2(2\pi - \Theta)r_+ $$

(12.46)

to the action. For a grand canonical ensemble, where the inverse temperature β and the rotational chemical potential μ are also held fixed, it is necessary to add further standard boundary terms at infinity. The final form of the Euclidean action is

$$ I = I_{can} - 2(2\pi - \Theta)r_+ + \beta M - \beta\mu J, $$

(12.47)

where I_{can} is the canonical action (2.12).

We can now evaluate the partition function (12.30) in the saddle point approximation. The classical BTZ solution has a time-independent metric, so the canonical action is zero, and the deficit angle Θ also vanishes. Thus

$$ Z[\beta, \mu] \approx \exp\{4\pi r_+ - \beta M + \beta\mu J\}, $$

(12.48)

where M, J, and r_+ are all to be viewed as functions of β and μ. Substituting our previous expressions for M, J, and r_+, we obtain

$$ \ln Z[\beta, \mu] \approx \frac{4\pi^2 \ell^2}{\beta(1 - \mu^2 \ell^2)}. $$

(12.49)

It is now easy to check that the thermodynamic relations

$$ \beta J = \frac{\partial \ln Z}{\partial \mu} $$

$$ M - \mu J = -\frac{\partial \ln Z}{\partial \beta} $$

$$ S = \ln Z + \beta(M - \mu J) $$

(12.50)

give back the mass, angular momentum, and entropy of the preceding section.

In contrast to the (3+1)-dimensional theory, one can also compute the first quantum correction to the path integral (12.30) – the Van Vleck–Morette determinant – by taking advantage of the Chern–Simons formalism. As discussed above, the topology of the Euclidean black hole spacetime is that of a solid torus, and we know from chapter 10 that the one-loop contribution to the path integral (12.30) is given in terms of the Ray–Singer torsion for this topology. This torsion has already been calculated, in the context of quantum cosmology, in reference [54]. A simple translation of notation gives, to the next order,

$$\ln Z \approx \frac{2\pi r_+}{4G} - \beta M + \beta \mu J + \frac{\pi r_+}{\ell}, \tag{12.51}$$

where I have restored the factors of G necessary to distinguish orders in the approximation.

12.4 Black hole statistical mechanics

Our discussion has focused so far on the thermodynamics of the (2+1)-dimensional black hole, and the outcome not been too different from the corresponding results in 3+1 dimensions. The simplicity of (2+1)-dimensional gravity offers us an additional opportunity, however: it may be possible to understand a microscopic 'statistical mechanical' source of black hole thermodynamics.

At first sight, this effort might seem futile. In ordinary statistical mechanics, entropy is a measure of the number of microscopic states that yield a fixed macroscopic state. But the entropy (12.29) can be arbitrarily large, while we have seen that the number of degrees of freedom of (2+1)-dimensional gravity is very small. Recall from section 6 of chapter 2, however, that the number of available degrees of freedom in a Chern–Simons theory increases drastically on a manifold with boundary: an infinite number of degrees of freedom that would ordinarily be discarded as 'pure gauge' become physical. These 'would-be gauge' degrees of freedom are natural candidates for the excitations that give rise to black hole entropy.

For this idea to work, we must first decide whether it is sensible to treat the horizon of a black hole as a boundary. The horizon is certainly not a genuine physical boundary, an edge of spacetime. It is, however, a place at which one must impose 'boundary conditions' in quantum gravity. Questions about black holes in quantum gravity are necessarily questions about conditional probabilities: for instance, 'If a black hole with a given mass and angular momentum is present, what is the probability of observing a certain spectrum of Hawking radiation?' To compute such probabilities, we must impose the appropriate conditions on the path integral; that is, we must restrict ourselves to paths for which the metric has properties that can be interpreted as meaning 'a black hole with the given mass and angular momentum is present'. The simplest way to do so is to split spacetime along a hypersurface as in figure 2.3

of chapter 2, and to compute separate path integrals inside and outside the surface, subject to appropriate boundary conditions that ensure that the surface is a horizon with the desired characteristics. But as we have seen, such a procedure requires additional boundary terms in the action, and leads to new dynamical degrees of freedom. In particular, for a negative cosmological constant $\Lambda = -1/\ell^2$ these degrees of freedom are described by the chiral WZW action (2.92) for $SL(2, \mathbf{R}) \times SL(2, \mathbf{R})$, or in the Euclidean theory, $SL(2, \mathbf{C})$.

We can now try to count the states of this WZW model. This is most easily done in the Euclidean theory [63]. Observe first that not all boundary states should be included. As we saw in chapter 2, many 'would-be diffeomorphism' degrees of freedom become dynamical, but those corresponding to Killing vectors do not; rather, Killing vectors generate a residual gauge invariance, and physical states must be invariant. For the Euclidean black hole, the Killing vectors correspond to rotations and time translations, and in the language of conformal field theory the corresponding condition on the states is that

$$L_0 |\text{phys}> = \bar{L}_0 |\text{phys}> = 0, \tag{12.52}$$

where the Virasoro operators L_0 and $\bar{L}_0$ generate the rigid displacements.

We can now appeal to the reasonably well-understood physics of the $SL(2, \mathbf{C})$ WZW model, which has been studied in the context of string theory and conformal field theory. If we let $\rho(N, \bar{N})$ denote the number of eigenstates of L_0 and $\bar{L}_0$ with eigenvalues N and $\bar{N}$, the partition function for this model is

$$Z_{SL(2,\mathbf{C})}(\tau)[A, \bar{A}] = Tr \left\{ e^{2\pi i \tau L_0} e^{-2\pi i \bar{\tau} \bar{L}_0} \right\} = \sum \rho(N, \bar{N}) e^{2\pi i \tau N} e^{-2\pi i \bar{\tau} \bar{N}}, \tag{12.53}$$

and the number of states satisfying (12.52), $\rho(0,0)$, can be obtained by a contour integral. But the partition function (12.53) is known from conformal field theory, at least for suitable values of the Chern–Simons coupling constant k. The details of the computation lie outside the scope of this book, but the final result is that to lowest order,

$$\rho(0, 0) \sim \exp \left\{ \frac{2\pi r_+}{4G} \right\}, \tag{12.54}$$

precisely the exponent of the Bekenstein–Hawking entropy (12.29). A similar result can be obtained by counting the states of the Lorentzian theory, although one must rely on some rather uncertain assumptions about the quantization of the more poorly understood $SL(2, \mathbf{R})$ WZW model [21, 60].

The horizon degrees of freedom that arise in the Chern–Simons approach to (2+1)-dimensional gravity thus provide a microscopic explanation for the entropy of the BTZ black hole. Unfortunately, it is difficult to generalize these results to 3+1 dimensions: as noted in chapter 2, similar degrees of freedom probably exist, but very little is known about their dynamics. The success of the (2+1)-dimensional model strongly suggests, however, that this avenue of research is worth pursuing.

13
Next steps

The universe in which we live is not (2+1)-dimensional, and the quantum theories described in this book are not realistic models of physics. Nor is (2+1)-dimensional quantum gravity fully understood; as I have tried to emphasize, many deep questions remain open. Nevertheless, the models developed in the preceding chapters can offer us some useful insights into realistic quantum gravity.

Perhaps the most important role of (2+1)-dimensional quantum gravity is as an 'existence theorem', a demonstration that general relativity can be quantized without any new ingredients. This is by no means trivial: there has long been a suspicion that quantum gravity would require a radical change in general relativity or quantum mechanics. While this may yet be true in 3+1 dimensions, the (2+1)-dimensional models suggest that no such revolutionary overhaul of known physics is needed. This does not mean that our existing frameworks are correct, of course, but it makes it less likely that major changes will come merely from the need to quantize gravity.

At the same time, (2+1)-dimensional quantum gravity serves as a sort of 'nonuniqueness theorem'. We have seen that there are many ways to quantize general relativity in 2+1 dimensions, and that not all of them lead to equivalent theories. This is perhaps not surprising, but it is a bit disappointing: in the absence of clear experimental tests of quantum gravity, there has been a widely held (although often unspoken) hope that the requirement of self-consistency might be enough to guide us to the correct formulation. The generalization to 3+1 dimensions is again risky, but the (2+1)-dimensional models suggest that we may not be so lucky. Of course, we do not yet know how much of the variation among models of (2+1)-dimensional gravity can be attributed to mere operator ordering ambiguities; I will return to this issue briefly below.

Beyond these generalities, (2+1)-dimensional quantum gravity casts some light on a number of the important conceptual problems of quantum gravity. A partial list is the following:

The 'problem of time': The covariant canonical quantization of chapter 6 gives a manifestly covariant quantum theory in a 'frozen time' formalism, in which no time-slicing is chosen and all observables are constants of motion. Nevertheless, we saw that it was possible to ask sensible questions about dynamics. This model is one of the few in which one can obtain an explicit realization of Rovelli's picture of 'evolving constants of motion' [230, 231]. Moreover, we saw that the model of chapter 6 was virtually equivalent to the manifestly dynamical, time-slice-dependent ADM quantum theory of chapter 5. The idea that 'frozen time' quantum gravity is the Heisenberg picture corresponding to a fixed-time-slicing Schrödinger picture is a central insight of (2+1)-dimensional gravity.

The problem of observables: A basic problem in quantum gravity arises from the fact that diffeomorphism-invariant observables are necessarily nonlocal. Without discounting the technical difficulties imposed by this fact, the (2+1)-dimensional models demonstrate that these difficulties are 'merely technical'. In particular, we have seen that nonlocal operators such as the modulus operator $\hat{\tau}$ or the holonomies $\hat{\mu}$ and $\hat{\lambda}$ can be used to fully reconstruct the spacetime geometry, classically and quantum mechanically. The extension to 3+1 dimensions is obviously difficult – in some sense, it requires the complete integration of the classical Einstein equations – but this seems to be a difficulty in practice, not in principle.

Nonperturbative quantization: We have seen that there is no need to assume a fixed background – and in particular, no requirement that a background be nearly flat – in order to quantize gravity. This is to be expected in canonical quantization, but even in the path integral methods of chapter 9, the 'background fields' $\bar{e}$ and $\bar{\omega}$ were ultimately integrated over, and played no fundamental role. The equivalence of the path integral and the canonical theory should reassure us that the use of background fields need not spoil nonperturbative outcomes.

Operator ordering: Not all of the quantum theories described in this book are equivalent, but many of the differences can be attributed to operator ordering. In retrospect, this is not surprising: ordering ambiguities are likely to be more severe than usual in a theory based on nonlocal observables. Even in ordinary quantum mechanics, such ambiguities can only be resolved by an appeal to experiment. The

(2+1)-dimensional models have demonstrated that these ambiguities are likely to be severe – the Hamiltonians of equation (5.38) have very different spectra, for instance – and unavoidable.

Topology change: Does consistent quantum gravity require spatial topology change? The answer in 2+1 dimensions is unequivocally no. The canonical quantum theories of chapters 5–7 are perfectly consistent descriptions of a universe with a fixed spatial topology. On the other hand, the path integrals of chapter 9 seem to allow the computation of amplitudes for tunneling from one topology to another. Problems with these topology-changing amplitudes remain, particularly in the regulation of divergent integrals over zero-modes. If these can be resolved, however, we will have to conclude that we have found genuinely different quantum theories of gravity.

Topology in quantum cosmology: In conventional descriptions of the Hartle–Hawking wave function, and in other Euclidean path integral descriptions of quantum cosmology, it is sometimes assumed that a few simple contributions dominate the sum over topologies. The results of chapter 10 indicate that such claims should be treated with skepticism; the sum in 2+1 dimensions seems to be dominated by an infinite number of complicated topologies, each individually exponentially suppressed. In four dimensions, this could lead to severe problems, since four-manifold topologies are not classifiable even in principle.* There may be ways around this difficulty, for instance by considering manifolds of bounded variation [48], but the problem is real.

Lessons for particular approaches: A number of concrete technical results for particular approaches to quantization have grown out of the corresponding (2+1)-dimensional models. For example, several properties of the loop operators $\widehat{T}^0$ and $\widehat{T}^1$ of chapter 7 were first discovered in 2+1 dimensions, and then adapted to 3+1 dimensions [11]. On a more pessimistic note, the difficulties we faced in chapter 8 have implications for the usefulness of the Wheeler–DeWitt equation in realistic (3+1)-dimensional gravity.

Realistic quantum gravity? Until recently, only one system in (2+1)-dimensional gravity has had direct relevance to realistic (3+1)-dimensional physics: the point particle solutions described briefly in chapters 1 and 3 have been used to draw conclusions about the behavior of cosmic strings and the problem of closed timelike curves [248]. The discovery of the (2+1)-dimensional black hole

---

* See, for example, reference [235].

has changed this. Black holes in 2+1 dimensions differ in important respects from their (3+1)-dimensional counterparts, but the similarities are significant enough to make the (2+1)-dimensional model quite valuable. In particular, while the statistical mechanical explanation of black hole entropy of chapter 12 has not yet been generalized to 3+1 dimensions, the (2+1)-dimensional model has certainly suggested a new and interesting place to look for an answer to one of the perennial questions of quantum gravity.

Where do we go from here? There are at least three broad areas in which research in (2+1)-dimensional quantum gravity should advance. We need to extend existing methods of quantization; we need to apply the results to new conceptual issues in quantum gravity; and we need to figure out how to incorporate nontrivial couplings with matter.

Much of this book has focused on 'quantum cosmologies' in which spatial slices have the topology of a torus. Perhaps the biggest open problem within the framework developed here is the extension of these results to more complicated topologies. The torus topology is atypical – it has an abelian fundamental group, for example, and an unusual mapping class group action. For conclusions based on spacetimes with the topology $[0,1] \times T^2$ to be fully trusted, they really ought to be checked in at least one other topology. This is a difficult problem, but the case of genus 2 seems within reach.

It may be that exact solutions are no longer possible in the case of more complicated topologies. If this is the case – and even if it is not – we need to develop suitable perturbation theories. Some preliminary steps in this direction have been taken in the classical context [214, 215], but quantum perturbation theory is almost completely unexplored. In particular, virtually nothing is known about perturbative approaches to the covariant canonical quantization of chapter 6.

I have tried to point out other open issues throughout the course of this book, ranging from the fundamental – 'solve the (2+1)-dimensional Wheeler–DeWitt equation' – to the more technical – 'work out the action of the mapping class group on the space of wave functions $\psi(r_2^+, r_2^-)$'. There is clearly no shortage of interesting questions.

Even within the limits of our present understanding, however, important conceptual problems have yet to be addressed. For example, what is the fate of a singularity in quantum gravity? An operator like the Hamiltonian $\hat{H}_{red}$ of equation (6.22) is clearly badly behaved at $T = 0$. On the other hand, Puzio has argued that in the ADM quantization of chapter 5, a wave packet initially concentrated away from the singular points in moduli space will remain nonsingular [222]. We do not really know how to ask the right quantum mechanical questions about singularities; Hosoya has proposed a quantum version of the *b*-boundary, for instance, but very little is yet understood about this construction [152].

Similarly, we do not really understand whether spatial and temporal intervals are quantized in (2+1)-dimensional quantum gravity. The results described at the end of chapter 7 are interesting but inconclusive, as is the analysis of 't Hooft's lattice approach in chapter 11. We do not know whether (2+1)-dimensional quantum gravity contains a natural mechanism for suppressing the cosmological constant, although Witten has shown that (2+1)-dimensional supersymmetry in the presence of gravity can ensure the vanishing of Λ without requiring the equality of bosonic and fermionic masses [31, 292]. We understand very little about the relationship between Euclidean and Lorentzian gravity in 2+1 dimensions, even though this model is a natural one in which to ask such questions.

Above all, we do not know how to incorporate matter into (2+1)-dimensional quantum gravity. Some limited progress has been made: there is evidence that scalar field theories coupled to (2+1)-dimensional gravity are renormalizable in the $1/N$ expansion [173, 201], certain matter couplings in supergravity have been studied [95, 191], and recent work on circularly symmetric 'midi-superspace models' has led to some interesting results [12, 16]. But the problem is difficult, in part because we lose the ability to represent diffeomorphisms as $ISO(2,1)$ gauge transformations.

Difficult as it is, however, an understanding of matter couplings may be the key to many of the conceptual issues of quantum gravity. One can explore the properties of a singularity, for example, by investigating the reaction of nearby matter, and one can look for quantization of time by examining the behavior of physical clocks. Moreover, many of the deep questions of quantum gravity can be answered only in the presence of matter. What is the final fate of an evaporating black hole? We can obtain a reliable answer only if we fully account for the quantum mechanics of Hawking radiation. Does gravity cut off ultraviolet divergences in quantum field theory? This idea is an old one [96, 161, 162], and it gets some support from the boundedness of the Hamiltonian in midi-superspace models [16], but it is only in the context of a full quantum field theory that a final answer can be given.

Finally, of course, the ultimate problem is to extend some or all of the methods of this book to 3+1 dimensions. I leave this as an exercise for the reader.

Appendix A
The topology of manifolds

This appendix provides a quick summary of the topology needed to understand some of the more complicated constructions in (2+1)-dimensional gravity. Readers familiar with manifold topology at the level of reference [192] or [204] will not learn much here, although this appendix may serve as a useful reference. The approaches I present here are not rigorous: this is 'physicists' topology', not 'mathematicians' topology', and the reader who wishes to pursue these topics further would be well advised to consult more specialized sources. A good intuitive introduction to basic concepts can be found in reference [164], and a very nice source for the visualization of two- and three-manifolds is reference [277].

Mathematically inclined readers may be somewhat surprised by my choice of topics. I discuss mapping class groups, for example, but I largely ignore homology. In addition, I introduce many concepts in rather narrow settings – for instance, I define the fundamental group only for manifolds. These choices represent limits of both space and purpose: rather than giving a comprehensive overview, I have tried merely to highlight the tools that have already proven valuable in (2+1)-dimensional gravity.

A.1 Homeomorphisms and diffeomorphisms

Let us begin by recalling the meaning of 'topology' in our context. Two spaces M and N are *homeomorphic* – written as $M \approx N$ – if there is an invertible mapping $f : M \to N$ such that

1. f is bijective, that is, both f and f^{-1} are one-to-one and onto; and

2. both f and f^{-1} are continuous.

Such a mapping is called a homeomorphism, and M and N are said to have the same topology. Roughly speaking, this means that M can

217

be continuously deformed into N, without any cutting or tearing; the homeomorphism f can be viewed as the instructions for this deformation.

If M and N are differentiable manifolds, the second condition on f can be strengthened to the requirement that f and f^{-1} be differentiable. In that case, M and N are said to be *diffeomorphic*, and f is called a diffeomorphism. In general, being diffeomorphic is a stronger condition than being homeomorphic, but in two and three dimensions, it can be shown that the two conditions are equivalent. This is no longer true in four dimensions, and the difference leads to the concept of 'exotic' differentiable structures [40].

A.2 The mapping class group

Valuable information about a manifold M can often be gained by looking at self-diffeomorphisms $f : M \to M$. Two such diffeomorphisms f_1 and f_2 are said to be *isotopic* if f_1 can be smoothly deformed into f_2, i.e., if there is a continuous family of diffeomorphisms

$$F(s) : M \to M \tag{A.1}$$

with $F(0) = f_1$ and $F(1) = f_2$. $F(s)$ is called an isotopy. Most of the diffeomorphisms that commonly arise in physics are isotopic to the identity map; that is to say, they can be built up from the identity by a series of infinitesimal diffeomorphisms. For topologically nontrivial manifolds, however, there often exist 'large' diffeomorphisms that are not isotopic to the identity.

The standard example of a large diffeomorphism is a *Dehn twist* of a two-dimensional torus T^2. This is the mapping from T^2 to itself obtained by cutting the torus along a circumference to form a cylinder, twisting one end by 2π, and gluing the ends back together (see figure A.1). Such a twist is a diffeomorphism – smooth curves are mapped to smooth curves, for instance – but it is fairly clear that it cannot be built up smoothly from infinitesimal deformations of T^2.

A *mapping class* is an equivalence class of diffeomorphisms of M, where f_1 is equivalent to f_2 if the two mappings are isotopic. The set of mapping classes of M naturally forms a group, the mapping class group, also called the homeotopy group. The mapping class group of an arbitrary two-surface is generated by Dehn twists around closed curves. In three or more dimensions much less is known, but some large diffeomorphisms can often be described quite explicitly [194]. Such transformations have been important in attempts to understand the spin and statistics of gravitational geons [114], and they play a vital role in the formulation of quantum gravity in 2+1 dimensions.

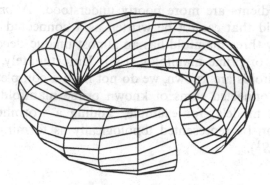

Fig. A.1.   A Dehn twist is the operation of cutting open a handle of a surface (in this case a torus) along a closed curve, twisting one end by 2π, and then regluing.

A.3 Connected sums

It is often useful to build a complicated manifold out of simpler building blocks. The ordinary (Cartesian) product is one such construction: if M and N are m- and n-dimensional manifolds, then $M \times N$ is an $(m + n)$-dimensional manifold. Many of the three-manifolds that occur in (2+1)-dimensional gravity have a product structure $[0, 1] \times \Sigma$, where $[0, 1]$ is a closed interval and Σ is a surface.

Manifolds of the same dimension can also be combined by means of *connected sums*. Let P and Q be connected n-dimensional manifolds, with points $p \in P$ and $q \in Q$. The connected sum $P\#Q$ is formed by deleting small n-dimensional balls around p and q and identifying the resulting spherical boundaries by an orientation-reversing map. The topology of $P\#Q$ can be shown to be independent of the choice of points p and q and the gluing map. It is not hard to see that the n-sphere is an 'identity element' for the connected sum, i.e., $M\#S^n \approx M$.

In two dimensions, orientable surfaces can be formed by taking connected sums of copies of the torus T^2. A *surface of genus g* is a connected sum of g copies of the torus,

$$\Sigma_g \approx \overbrace{T^2\# \cdots \#T^2}^{g\ times},$$ (A.2)

i.e., a 'g-holed donut'. Such a surface is said to have g *handles*. A basic result in the classification of surfaces is that any compact orientable two-manifold is homeomorphic to either a two-sphere or some surface of genus $g \geq 1$.

In three dimensions, a similar decomposition theorem holds, although the basic ingredients are more poorly understood. A *prime manifold* is a three-manifold that cannot be written as a connected sum. It can be shown that any three-manifold has an almost unique decomposition as a connected sum of prime manifolds [146]. Unfortunately, while many of the prime manifolds are known, we do not have a complete classification. Nevertheless, connected sums of known prime manifolds often provide useful physical models; for example, 'adding a wormhole' to a three-manifold M can be represented topologically as forming the connected sum $M\#(S^2 \times S^1)$.

A.4 The fundamental group

To characterize the topology of a manifold M, we must find topological invariants, quantities associated with M that are invariant under homeomorphisms. One such topological invariant is the *fundamental group* $\pi_1(M)$, also known as the *first homotopy group* of M. Roughly speaking, the fundamental group counts the number of 'topologically distinct' closed curves on M. On the plane, for instance, any loop can be deformed into any other without breaking, and the fundamental group has only one element. On the annulus, in contrast, distinct loops can be characterized by a winding number, and the fundamental group is isomorphic to the group of integers.

Let us make these notions more precise. A curve from x_0 to x_1 in M can be represented as a continuous mapping

$$\gamma : [0,1] \to M, \qquad \gamma(0) = x_0, \qquad \gamma(1) = x_1. \qquad (A.3)$$

A loop based at x_0 is a curve with $\gamma(0) = \gamma(1) = x_0$. Two loops based at x_0 are said to be *homotopic* if one can be continuously deformed into the other without moving the base point; that is, $\gamma_1 \sim \gamma_2$ if there is a continuous map

$$F : [0,1] \times [0,1] \to M \qquad (A.4)$$

such that

$$F(t,0) = \gamma_1(t)$$
$$F(t,1) = \gamma_2(t)$$
$$F(0,s) = F(1,s) = x_0. \qquad (A.5)$$

It is easy to check that homotopy is an equivalence relation; if $\gamma_1 \sim \gamma_2$ and $\gamma_2 \sim \gamma_3$, then $\gamma_1 \sim \gamma_3$. We denote the class of loops homotopic to γ by $[\gamma]$, and the set of all homotopy classes by $\pi_1(M, x_0)$.

On the plane, we can always choose an interpolating function

$$F(t, s) = s\gamma_2(t) + (1 - s)\gamma_1(t). \tag{A.6}$$

This provides a homotopy between any two loops γ_1 and γ_2, and the plane thus has only one homotopy class. On the annulus, in contrast, such a simple construction will fail: for some values of s, the image of F will typically lie outside the annulus. This is not surprising, since we should not expect to be able to deform a loop around the center of an annulus to one that does not enclose the central hole. Instead, it may be shown that any two loops with the same winding number are homotopic, and that the homotopy classes of the annulus are labeled by the integers.

To make $\pi_1(M, x_0)$ into a group, we must define a product, a unit element, and an inverse. To obtain a product of homotopy classes, we first construct a product of individual loops. The product of two loops γ_1 and γ_2 is the loop obtained by going first around γ_1 and then around γ_2,

$$\gamma_1 \cdot \gamma_2(t) = \begin{cases} \gamma_1(2t) & \text{if } t \le 1/2 \\ \gamma_2(2t - 1) & \text{if } t > 1/2 \end{cases} \tag{A.7}$$

For such a product of individual loops to make sense as a product of homotopy classes, the result must be independent of which loop we choose to represent each equivalence class,

$$\alpha \cdot \gamma \sim \beta \cdot \delta \quad \text{if} \quad \alpha \sim \beta \quad \text{and} \quad \gamma \sim \delta. \tag{A.8}$$

Fortunately, this is true, and we can unambiguously define a product on $\pi_1(M, x_0)$ by

$$[\gamma_1] \cdot [\gamma_2] = [\gamma_1 \cdot \gamma_2]. \tag{A.9}$$

For the annulus, for instance, this product corresponds to ordinary addition, since the result of attaching a loop with winding number m to one with winding number n is a loop with winding number $m + n$. For more complicated manifolds, this product need not be commutative.

The unit element for this product is the homotopy class of the constant path $c : [0, 1] \mapsto x_0$. A loop is thus homotopic to the identity if it can be shrunk to the base point x_0. Such a loop is said to be *contractible*.

To define the inverse of a homotopy class, we take the 'inverse' of a loop γ to be the loop obtained by traversing γ in the opposite direction,

$$\gamma^{-1}(t) = \gamma(1 - t). \tag{A.10}$$

This is not really an inverse, since $\gamma_1 \cdot \gamma_1^{-1} \neq c$, but it is not hard to prove that the product $\gamma_1 \cdot \gamma_1^{-1}$ can be *deformed* to the constant map, so

$$[\gamma] \cdot [\gamma^{-1}] = [c]. \tag{A.11}$$

Once again, this definition has an obvious significance on the annulus: the inverse of a loop with winding number n is a loop with winding number $-n$, and the product of the two can be unwound and shrunk to a point.

It is not hard to show that $\pi_1(M, x_0)$ is indeed a group. From its construction, this group depends on both the manifold M and the base point x_0. If M is connected, however, the base point dependence is largely illusory: for any two points x_0 and x_1, $\pi_1(M, x_0)$ and $\pi_1(M, x_1)$ are isomorphic. In practice, it is therefore common to drop the reference to the base point and to refer to the fundamental group $\pi_1(M)$.

A manifold M is said to be *simply connected* if all loops in M are homotopic to the identity. The fundamental group of a simply connected manifold is the trivial group, consisting only of the unit element. A standard example of a simply connected manifold is the n-dimensional sphere S^n for $n > 1$. (This is the theorem that 'you can't lasso a sphere'.)

The importance of the fundamental group comes from the fact that it is a topological invariant: if M is homeomorphic to N, then $\pi_1(M)$ is isomorphic to $\pi_1(N)$. For two-dimensional surfaces, in fact, $\pi_1(M)$ completely characterizes the topology: any two surfaces with isomorphic fundamental groups are homeomorphic. In more than two dimensions this is no longer true, but the fundamental group still gives an important partial characterization of the topology.

A.5 Covering spaces

The fundamental group is essentially a description of the way closed curves wind around a manifold. It is often useful to 'unwrap' these curves, associating with the original manifold a 'larger' manifold with fewer noncontractible curves.

A *covering space* of M is a new manifold that looks locally like a discrete set of copies of M. More precisely, a covering space of M is a manifold N, together with a projection $p : N \to M$ such that

1. p is surjective (onto) and continuous; and

2. every point $x \in M$ has a neighborhood U such that $p^{-1}(U)$ is homeomorphic to $U \times \Lambda$, where Λ is a discrete space.

The cardinality of $p^{-1}(x)$ is called the multiplicity of the covering at the point x, and a point $y \in p^{-1}(x)$ is said to lie above x. If the multiplicity is a constant n, we speak of an n-fold covering. Figure A.2, for instance, shows a two-fold covering of an annulus.

A path γ in M is said to *lift* to a path $\tilde{\gamma}$ in N if $p \circ \tilde{\gamma} = \gamma$. The lift $\tilde{\gamma}$ is not quite unique: for an n-fold covering, the initial point $\gamma(0) = x$ of γ can be lifted to any one of the n points $y_i \in p^{-1}(x)$ lying above x. But

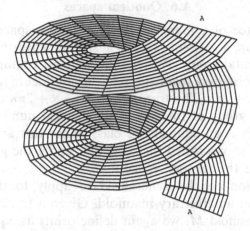

Fig. A.2. A double covering of the annulus is obtained by identifying the two edges (labeled *A*) of this figure.

this is the only source of ambiguity; once the starting point $\tilde{\gamma}(0)$ is fixed, $\tilde{\gamma}$ is determined uniquely.

Notice that in figure A.2, a loop with winding number two on the annulus lifts to a loop with winding number one in the covering space, while loop with winding number one lifts to an open path. Since the lifting process can 'unwrap' a loop, we should expect that a covering space will have a simpler fundamental group than the original manifold. Moreover, by choosing different covering spaces we can unwrap different sets of loops, so we might expect to find a variety of different ways of simplifying $\pi_1(M)$.

This intuition is correct. If N is a covering space of M with projection $p : N \rightarrow M$, p can be used to project loops from N to M. This projection induces a group homomorphism $p_* : \pi_1(N, y) \rightarrow \pi_1(M, p(y))$ between fundamental groups, which is one-to-one but not necessarily onto. Consequently, the image of $\pi_1(N)$ under p_* is a subgroup of $\pi_1(M)$, called the characteristic subgroup of the covering space. The characteristic subgroup essentially consists of those noncontractible loops in M that lift to loops rather than open paths in N.

If M is connected, it turns out that *any* subgroup of $\pi_1(M)$ is the characteristic subgroup of *some* covering space. In other words, given any subgroup G of $\pi_1(M)$, we can find a covering space in which all of the loops of $\pi_1(M)$ are 'unwrapped' except for those in G. In particular, the trivial group $\{1\}$ is always a subgroup of $\pi_1(M)$, so there must be a covering space whose fundamental group is trivial. This simply connected covering space is called the *universal cover* of M, and is denoted by $\widetilde{M}$.

A.6 Quotient spaces

If a covering space of a manifold M is a 'larger' space with a simpler fundamental group, a *quotient space* is a 'smaller' space with a more complicated fundamental group. Perhaps the simplest example of a quotient space is a circle obtained from the real line by identifying each point x with the points $x + 2\pi n$. The translations $x \mapsto x + 2\pi n$ can be viewed as an action of the group of integers $\mathbf{Z}$ on the real numbers $\mathbf{R}$, and each point in the circle is an equivalence class under this action. The set of points equivalent to x is called the *orbit* of x under the group action, and the set of orbits is the quotient space $\mathbf{R}/\mathbf{Z}$.

This construction can be generalized to apply to the action of an arbitrary group on an arbitrary manifold. Given a transformation group G acting on a manifold M, we again define orbits as equivalence classes under the group action, and write the space of orbits as M/G. There is a natural projection $p : M \to M/G$ of points in M to their orbits; we fix the point set topology of M/G by defining open sets in M/G to be those sets U for which $p^{-1}(U)$ is open in M.

In general, a quotient space M/G defined in this way will not be a manifold. For instance, let M be a circle and let G be the group generated by a rotation by an irrational angle. It is easy to see that a typical orbit will be a dense set of points in the circle, and it is intuitively clear that the quotient space will not be very well-behaved (in fact, it won't even be Hausdorff). If we want M/G to have nice topological properties, we will need additional conditions on the group action.

The appropriate requirements are that the action of G be *free* and *properly discontinuous*. An action of G on M is free if the isotropy group $G_x = \{g \in G : gx = x\}$ is trivial at every point, that is, $gx \neq x$ unless g is the identity. An action is properly discontinuous if

1. any two points x and x' in M that do not lie on the same orbit have neighborhoods U and U' such that $gU \cap U' = \emptyset$ for all $g \in G$;

2. for each $x \in M$, the isotropy group G_x is finite; and

3. every point $x \in M$ has a neighborhood U such that $gU = U$ for $g \in G_x$, and $gU \cap U = \emptyset$ for $g \notin G_x$.

In essence, these conditions ensure that an open subset of M will overlap with only a finite number of its images under G, thus avoiding the obvious pathologies that could occur in the quotient space.

If G acts freely and properly discontinuously on a manifold M, it may be shown that the quotient M/G is a manifold. If the action is properly discontinuous but not free, M/G is an *orbifold*, essentially a manifold with certain mild singularities. A typical orbifold is the quotient of the plane by the group generated by a $2\pi/n$ rotation: the action is not free at the origin, and the quotient is a manifold with a conical singularity.

Each point in the quotient M/G is an orbit in M, and each such orbit may be labeled by a giving a point $p \in M$ lying on the orbit. A complete set F of such representatives (or its closure) with a suitable topology is called a *fundamental region* for the action of G on M.

In general, the topology of a quotient M/G will be quite complicated. But if M is simply connected and if G is a finite group acting freely on M, it may be shown that

$$\pi_1(M/G) \approx G. \qquad (A.12)$$

In that case, we can relate the quotient construction to our earlier discussion of covering spaces: the projection $p : M \to M/G$ is a covering map, and M is the universal cover of M/G. The circle $\mathbf{R}/\mathbf{Z}$ again provides a nice example of this phenomenon: for $M = \mathbf{R}$ and $G = \mathbf{Z}$, it is indeed true that $\pi_1(\mathbf{R}/\mathbf{Z}) = \pi_1(S^1) \approx \mathbf{Z}$.

It turns out that a sort of converse of (A.12) is also true: any connected manifold M can be written as a quotient space

$$M \approx \widetilde{M}/\pi_1(M), \qquad (A.13)$$

where $\widetilde{M}$ is the universal covering space of M. For this expression to make sense, we must specify an action of $\pi_1(M)$ on $\widetilde{M}$. The basic idea is to unwrap loops in M into open paths in $\widetilde{M}$, and to map the starting point of each path to its endpoint. More precisely, let γ be a noncontractible loop in M based at x, representing an element $[\gamma] \in \pi_1(M, x)$, and let y_1 be a point in $\widetilde{M}$ lying over x (i.e., $p(y_1) = x$). We saw in the last section that γ lifts to a unique path $\tilde{\gamma}$ in $\widetilde{M}$ starting at y_1. Since $\widetilde{M}$ is simply connected, $\tilde{\gamma}$ will be an open path, with an endpoint $\tilde{\gamma}(1) = y_2$ that also lies over x. We define a map D_γ by $y_2 = D_\gamma y_1$.

Now consider any other point $x' \in M$ different from the basepoint x. Choose a path α from x' to x, and recall that α^{-1} denotes the reverse path from x to x', so the composite path $\alpha^{-1} \circ \gamma \circ \alpha$ is a closed loop based at x'. We can now construct a map $D_{\alpha^{-1} \circ \gamma \circ \alpha}$ exactly as above. It can be shown that D is actually independent of the particular path α, and depends only on the homotopy class of γ. We have therefore defined a transformation $D_{[\gamma]} : \widetilde{M} \to \widetilde{M}$ for each $[\gamma] \in \pi_1(M)$, which by its construction satisfies $p \circ D_{[\gamma]} = p$. Such a transformation is called a *deck transformation* or *covering transformation* of $\widetilde{M}$.

It is straightforward to check that

$$D_{[\gamma_1]} \circ D_{[\gamma_2]} = D_{[\gamma_1] \cdot [\gamma_2]}, \qquad (A.14)$$

so the deck transformations give an action of $\pi_1(M)$ on $\widetilde{M}$. This action can be shown to be free and properly discontinuous. It is then evident that the quotient $\widetilde{M}/\pi_1(M)$ determined by this action identifies all of the points $y_i \in \widetilde{M}$ lying above any given point $x \in M$, thus collapsing $\widetilde{M}$ down to a single copy of M, verifying (A.13).

A.7 Geometrization

Quotient spaces as we have defined them are topological constructs, but it is often possible to add additional structure. For example, if M has a Riemannian metric and G is a group of isometries of that metric, then M/G will inherit a metric; if M is a complex manifold and G acts holomorphically, M/G will inherit a complex structure.

In two dimensions, one of the fundamental results of Riemann surface theory, the *uniformization theorem*, makes use of this fact. Let $\widehat{\mathbf{C}} = \mathbf{C} \cup \infty$ be the Riemann sphere, with its standard metric of constant curvature $+1$,

$$ds^2 = \frac{4dzd\bar{z}}{(1+|z|^2)^2};\tag{A.15}$$

let $\mathbf{C}$ be the complex plane, with its standard metric of constant curvature 0,

$$ds^2 = dzd\bar{z};\tag{A.16}$$

and let $\mathbf{H}^2 = \{z \in \mathbf{C} : Imz > 0\}$ be the upper half-plane, with its standard metric of constant curvature -1,

$$ds^2 = \frac{dzd\bar{z}}{(Imz)^2}.\tag{A.17}$$

It can be shown that every surface is homeomorphic to D/Γ, where D is either $\widehat{\mathbf{C}}$, $\mathbf{C}$, or $\mathbf{H}^2$ and Γ is a properly discontinuous group of isometries of the constant curvature metric on D. This means that every surface admits a constant curvature metric, and that the classification of such metrics may be reduced to an algebraic problem of classifying groups of isometries of a few covering spaces. We thus obtain a fruitful connection between topology, geometry, and algebra.

An important open question is whether this idea can be extended to three-dimensional manifolds. As in two dimensions, there are a small number of 'geometric' three-manifolds that can occur as covering spaces. Thurston's geometrization conjecture is that any three-manifold can be decomposed into pieces that are quotient spaces of one of these geometric

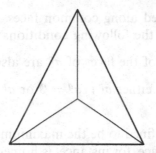

Fig. A.3. A triangle is subdivided by inserting an additional vertex and three additional edges. See figure 11.3 for the subdivision of a tetrahedron.

manifolds (the exact definition of 'decomposition' is a bit complicated). This conjecture has been shown to hold in a large number of special cases, but a general proof is still lacking [257].

A.8 Simplices and Euler numbers

To understand (2+1)-dimensional gravity, we shall need one more key topological invariant, the Euler number of a manifold. Let us start with a simple example, a two-dimensional surface. Any surface Σ can be *triangulated*, that is, broken up into a set of triangles joined along their edges (see figure 11.1). Given such a triangulation, the *Euler number* χ of Σ is defined as

$$\chi(\Sigma) = F - E + V, \tag{A.18}$$

where F is the number of faces, E is the number of edges, and V is the number of vertices in the triangulation.

It is not too hard to show that $\chi(\Sigma)$ depends only on the surface Σ, and not on the triangulation. This can be made plausible by considering the effect of subdividing a triangle (see figure A.3): the subdivision introduces two new faces, three new edges, and one new vertex, thus leaving χ unchanged. It is also clear that $\chi(\Sigma)$ is a topological invariant of Σ, that is, that it is preserved by homeomorphisms.

To generalize this construction to higher dimensions, we first need the p-dimensional generalization of a triangle. Given a set of $p + 1$ linearly independent vertices $\{x_1, \ldots, x_{p+1}\} \in \mathbf{R}^n$, a *p-simplex* σ^p is the set

$$\sigma^p = \left\{ x = \sum_{i=1}^{p+1} \lambda_i x_i : \lambda_i \geq 0, \sum_{i=1}^{p+1} \lambda_i = 1 \right\}. \tag{A.19}$$

The jth face of σ^p is the subsimplex $\{x \in \sigma^p : \lambda_j = 0\}$. A *simplicial complex* is a set of simplices joined along common faces, or more precisely a set K of simplices satisfying the following conditions:

1. If $\sigma^p \in K$, then all of the faces of σ^p are also in K; and

2. If $\sigma^p, \sigma^q \in K$, then either $\sigma^p \cap \sigma^q = \emptyset$ or $\sigma^p \cap \sigma^q$ is a common face of σ^p and σ^q.

The dimension of K is defined to be the maximum dimension of a simplex in K. A triangulated surface, for instance, is homeomorphic to a simplicial complex of dimension 2; a 3-simplex is a solid tetrahedron.

The Euler number of an m-dimensional simplicial complex K is defined in exact analogy to the two-dimensional case:

$$\chi(K) = \sum_{p=0}^{m} (-1)^p N_p, \tag{A.20}$$

where N_p is the number of p-simplices σ^p in K. $\chi(K)$ can again be shown to be a topological invariant, and can therefore be defined for any manifold homeomorphic to a simplicial complex. An important result is that $\chi(M) = 0$ for any closed odd-dimensional manifold.

For dimension greater than two, the Euler number gives only limited information about the topology. More powerful topological invariants, the *homology groups*, can be built from simplicial complexes, but I shall not discuss these here; see, for instance, [192] or [204]. It is worth mentioning that the choice of dividing a manifold into simplices instead of other building blocks is somewhat arbitrary; for instance, a 'cell decomposition' of a M into p-dimensional balls can also be used to determine the Euler number and the homology groups.

The Euler number is important in part because of the *Poincaré–Hopf index theorem*, which relates $\chi(M)$ to the properties of vector fields on M. Let v be a vector field with a zero at some point $x \in M$. The *index* of v at x is a measure of the directional change of v around x, determined as follows. Choose a neighborhood of x homeomorphic to $\mathbf{R}^n$, and consider a small $(n-1)$-sphere S_ϵ around x. The unit vector $v/|v|$ on S_ϵ can then be viewed as a map from S_ϵ to the unit sphere, and the index of v at x is defined as the degree of this map.* In two dimensions, the index measures the number of times v rotates as we move counterclockwise around x; a source has index 1, while a saddle has index -1.

The Poincaré–Hopf theorem states that if v is any smooth vector field with isolated zeros on a compact oriented manifold M, and if v points

---

* The degree is the number of times S_ϵ covers the unit sphere; for a precise definition, see reference [76], [192], or [196].

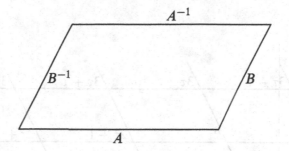

Fig. A.4.   A torus may be cut open to form a parallelogram.

outward at each boundary component of M, then [196]

$$ind(v) = \chi(M), \qquad (A.21)$$

where $ind(v)$ denotes the sum of the indices of v at all of its zeros. In particular, a vector field on a compact manifold with $\chi(M) \neq 0$ must have zeros. (This is the theorem that 'you can't comb a hairy billiard ball without leaving a bald spot'.) Conversely, a vector field with $ind(v) = 0$ can always be deformed to one that vanishes nowhere on M. This will be important in appendix B when we discuss the existence of Lorentzian metrics on $(2+1)$-dimensional spacetimes.

A.9 An application: topology of surfaces

Let us now apply these ideas to the topology of two-dimensional orientable surfaces. As noted above, any such surface is homeomorphic to either a sphere S^2 or the connected sum of g tori. The number g is the genus; the sphere is said to have genus zero.

It is clear that the two-sphere has a trivial fundamental group, that is, that any loop on S^2 is contractible. S^2 is thus its own universal covering space. To obtain its Euler number, note that S^2 is homeomorphic to the surface of a tetrahedron. This surface is a simplicial complex with four faces, six edges, and four vertices, so by (A.18), $\chi(S^2) = 2$.

Next consider the torus T^2. Let us cut the surface open to form a parallelogram P (figure A.4), with edges labeled A, B, A^{-1}, and B^{-1}; to recover the original torus, we sew together opposite edges, identifying A with A^{-1} and B with B^{-1}. It is evident from the figure that A is a noncontractible curve on T^2, and thus represents an element of $\pi_1(T^2)$. A^{-1} represents the same curve traversed in the opposite direction, so as a homotopy class it is indeed the inverse of A. Similarly, B and B^{-1} represent another element of $\pi_1(T^2)$ and its inverse.

The product AB is a spiral going once around the torus. Other noncontractible curves can be obtained by taking more complicated products, and

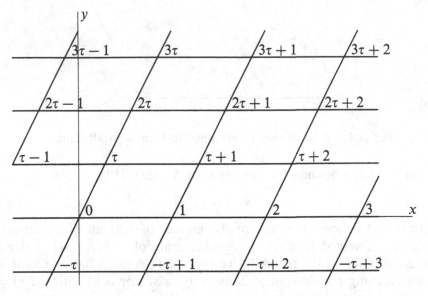

Fig. A.5.   The torus of figure A.4 can be duplicated to form a tiling of the plane. Corresponding points in any two tori are related by transformations (A.24).

in fact, any noncontractible curve on T^2 is homotopic to some product of this type; that is, A and B generate $\pi_1(T^2)$. These generators are not independent, however: the curve $ABA^{-1}B^{-1}$, which encircles the parallelogram in figure A.4, can clearly be shrunk to a point. The fundamental group of the torus therefore has two generators and one relation,

$$\pi_1(T^2) = \left\langle A, B \,|\, ABA^{-1}B^{-1} = 1 \right\rangle. \tag{A.22}$$

The relation simply means that the two generators commute, so $\pi_1(T^2)$ is a free abelian group on two generators,

$$\pi_1(T^2) \approx \mathbf{Z} \oplus \mathbf{Z}. \tag{A.23}$$

To obtain the universal cover of T^2, first place the parallelogram P of figure A.4 on the complex plane, with vertices 0, 1, τ, and $\tau + 1$, where τ is a complex number with positive imaginary part, the modulus. By duplicating this figure, we obtain a 'tiling', or *tessellation*, of the plane, as in figure A.5. We can move from a point on one parallelogram to the corresponding point on another by means of a transformation of the form

$$z \mapsto z + m + n\tau. \tag{A.24}$$

These transformations describe a free, properly discontinuous action of $\mathbf{Z} \oplus \mathbf{Z}$ on $\mathbf{C}$, and we can form the quotient space $\mathbf{C}/\mathbf{Z} \oplus \mathbf{Z}$. The transformation $z \mapsto z + 1$ identifies the edges A and A^{-1} of P, while $z \mapsto z + \tau$ identifies B and B^{-1}, so the quotient space is simply the original torus, and $\mathbf{C}$ can be identified as the universal cover of T^2. Each parallelogram in figure A.4 is a fundamental region for the group action.

This quotient space representation allows us to see the action of the mapping class group of T^2 – also known as the modular group – quite explicitly. Consider a Dehn twist around the curve B – that is, cut the torus open along B, twist by 2π, and reglue the edges. This process will clearly not change B, but A will be twisted into a spiral, $A \mapsto AB$. The effect on the tiling of the plane will thus be to shift τ to $\tau + 1$. A Dehn twist around A is a bit harder to visualize, but its effect is to transform τ to $\tau/(\tau + 1)$. These two transformations generate the mapping class group of the torus; a more common, but equivalent, representation is given by

$$S : \tau \to -\frac{1}{\tau}$$
$$T : \tau \to \tau + 1. \tag{A.25}$$

We next consider the geometry corresponding to this quotient construction. A torus $\mathbf{C}/\mathbf{Z} \oplus \mathbf{Z}$ will inherit a flat metric and a complex structure from the plane. Tori obtained from parallelograms with different values of τ will all be homeomorphic, of course – any torus is homeomorphic to any other – but in general, the homeomorphisms will not be isometries. Hence different values of τ will ordinarily label different flat metrics on T^2. The resulting parameter space $\{\tau \in \mathbf{C} : Im\tau > 0\}$ is called the *Teichmüller space $\mathcal{N}$* of the torus.

Not all values of τ give distinct geometries, however. We have seen that diffeomorphisms in the mapping class group, which surely cannot change the metric, do change τ. If we identify the values of τ that differ by transformations of the form (A.25), we obtain the true parameter space of flat metrics on the torus modulo diffeomorphisms, the *moduli space*. Points in moduli space are orbits of the mapping class group, which acts properly discontinuously on Teichmüller space; if we denote the mapping class group by $\mathcal{D}$, the torus moduli space is the quotient space $\mathcal{N} = \mathcal{N}/\mathcal{D}$. A fundamental region for the group action is the 'keyhole' region shown in figure A.6.

Finally, to obtain the Euler number of the torus we must find a triangulation of T^2. A suitable triangulation is shown in figure A.7, and by counting simplices we see that $\chi(T^2) = 0$.

We are now ready to tackle higher genus surfaces Σ_g. Just as a torus can be cut open into a parallelogram, a genus g surface can be dissected

Appendix A

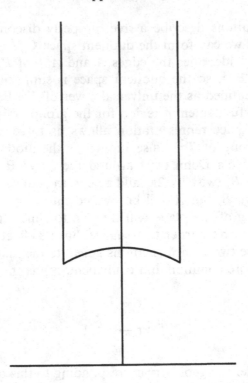

Fig. A.6. A fundamental region for the action of the modular group (A.25) on the upper half-plane is given by the region $-1/2 \leq \tau \leq 1/2$, $|\tau| \geq 1$. Points $\tau = -1/2 + i\tau_2$ and $\tau = 1/2 + i\tau_2$ must be identified in this figure, as must points $\tau = \exp\{i\theta\}$ and $\tau = \exp\{-i\theta\}$.

into a $4g$-sided polygon, with alternating edges identified as in figure A.8. It is not hard to see that sets of four adjacent edges will glue back together to give 'handles' of Σ_g. The sides of this $4g$-gon are again noncontractible curves, and represent elements of the fundamental group of Σ_g. As in the case of the torus, there is one relation among these generators, since the path completely surrounding the $4g$-gon can be shrunk to a point. Hence

$$\pi_1(\Sigma_g)$$
$$= \left\langle A_1, B_1, \ldots, A_g, B_g \,|\, A_1 B_1 A_1^{-1} B_1^{-1} \cdots A_g B_g A_g^{-1} B_g^{-1} = 1 \right\rangle. \quad \text{(A.26)}$$

A group of this form is called a *surface group*. Note that for genus greater than one, $\pi_1(\Sigma_g)$ will no longer be abelian.

To obtain the universal cover of Σ_g, we would again like to tile the plane with $4g$-gons. This is not possible on the flat plane. However, it may be shown that appropriate tessellations do exist on the hyperbolic plane $\mathbf{H}^2$,

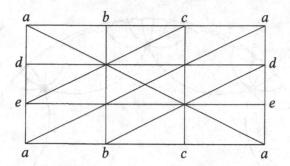

Fig. A.7. To obtain a torus, one must identify opposite edges of this diagram. Vertices labeled by the same letter are therefore identified, and it is easy to see that the diagram contains 20 faces, 30 edges, and 10 vertices.

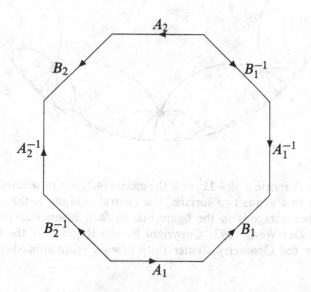

Fig. A.8. A genus two surface can be cut open to form an octagon, with edges identified as shown. (See also figure 4.1.)

as illustrated in figure A.9. The corresponding transformation groups can be shown to be discrete subgroups of $PSL(2, \mathbf{R})$, the group of isometries of the constant negative curvature metric (A.17). These subgroups again act properly discontinuously, and we can write

$$\Sigma_g \approx \mathbf{H}^2 / \Gamma, \tag{A.27}$$

where Γ is a representation of $\pi_1(\Sigma_g)$ in $PSL(2, \mathbf{R})$. Like the torus, the surface Σ_g inherits a metric – now with constant negative curvature – from its covering space.

The surface Σ_g admits a group of large diffeomorphisms, generated by

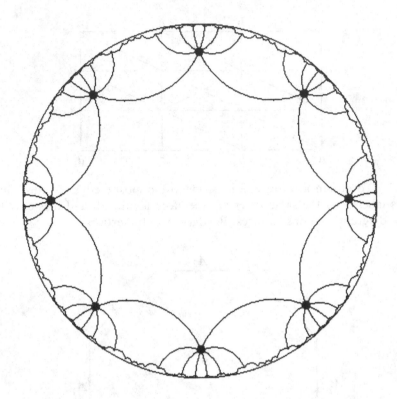

Fig. A.9.   The Poincaré disk – $\mathbf{H}^2$ with the metric (4.2) – is tessellated by octagons corresponding to a genus two surface. The central octagon is that of figure 4.1; it and the other octagons in the figure are, in fact, isometric. [Image created by Deva Van Der Werf, 1993. Copyright by the Regents of the University of Minnesota for the Geometry Center (http://www.geom.umn.edu). Used with permission.]

Dehn twists around noncontractible curves. This mapping class group $\mathcal{D}_g$ will again act on the tiling of $\mathbf{H}^2$ by Σ_g by mixing the generators of $\pi_1(\Sigma_g)$. It may be shown that this action gives precisely the outer automorphisms[†] of $\pi_1(\Sigma_g)$ [37]. Now, however, it is much harder to express these transformations explicitly in terms of changes in the tiling, and no simple analog of equation (A.25) is known. We can again define the genus g Teichmüller space $\widetilde{\mathcal{N}}_g$ as the space of tilings, and the moduli space $\mathcal{N}_g = \widetilde{\mathcal{N}}_g/\mathcal{D}_g$ as the space of inequivalent constant negative curvature metrics, but an explicit parameterization becomes much more difficult.

   Finally, let us compute the Euler number of Σ_g. The simplest approach

---

[†] The inner automorphisms $\mathrm{Inn}(G)$ of a group G are the transformations of the form $g \mapsto hgh^{-1}$; the outer automorphisms are the equivalence classes $\mathrm{Aut}(G)/\mathrm{Inn}(G)$.

is to take advantage of the fact that $\Sigma_g \approx T^2 \# \Sigma_{g-1}$, where $\#$ denotes the connected sum. Given triangulations of Σ_{g-1} and T^2, we can form the required connected sum by removing one face from each surface and identifying three edges and three vertices. The lost edges and vertices cancel in the Euler number (A.18), but the total number of faces is reduced by two. Thus

$$\chi(T^2 \# \Sigma_{g-1}) = \chi(\Sigma_{g-1}) + \chi(T^2) - 2, \tag{A.28}$$

and using our previous result that $\chi(T^2) = 0$, we see that

$$\chi(\Sigma_g) = 2 - 2g. \tag{A.29}$$

Appendix B
Lorentzian metrics and causal structure

In general relativity we are interested in both the topology and the geometry of spacetime. The body of this book concentrates on geometrical issues in (2+1)-dimensional gravity and their physical implications, while appendix A introduces some basic topological concepts. The purpose of this appendix is to briefly discuss a set of issues intermediate between topology and geometry: issues of the large scale structure, and in particular the causal structure, of a spacetime with a Lorentzian metric.

Questions of large scale structure have played a very important role in recent work in (3+1)-dimensional general relativity, leading to general theorems about singularities, causality, and topology change. A thorough discussion is given in reference [145] (see also [123]). Many of these general results have not yet been applied to 2+1 dimensions, and I shall not attempt to review them here; my aim is merely to introduce the ideas that have already found a use in (2+1)-dimensional gravity.

B.1 Lorentzian metrics

To specify a spacetime, we need a manifold M with a Lorentzian metric, that is (in three dimensions) a metric g of signature $(-++)$. Such a metric determines a light cone at each point in M. A spacetime M is *time-orientable* if a continuous choice of the future light cone can be made, that is, if there is a global distinction between the past and future directions. Similarly, M is *space-orientable* if there is a global distinction between left- and right-handed spatial coordinate frames.

A *line element field*, or *direction field*, is a continuous choice of an unordered pair $(n, -n)$ of nowhere vanishing vectors on M. It is not hard to show that a paracompact manifold M will admit a Lorentzian metric if and only if it admits a line element field. Indeed, if g is a Lorentzian metric on M, we can choose an arbitrary auxiliary Riemannian (positive definite) metric h – such a metric always exists if M is paracompact – and consider the eigenvalue equation

$$g_{\mu\nu}n^\nu = \lambda h_{\mu\nu}n^\nu. \tag{B.1}$$

This equation will have one negative eigenvalue, with a corresponding eigenvector that can be normalized by setting

$$g_{\mu\nu}n^\mu n^\nu = -1, \tag{B.2}$$

which fixes n up to overall sign. Thus $(n, -n)$ will be a line element field. Conversely, if M admits a line element field $(n, -n)$, we can again choose a Riemannian metric h, normalized so that

$$h_{\mu\nu}n^\mu n^\nu = 1, \tag{B.3}$$

and set

$$g_{\mu\nu} = -2(h_{\mu\lambda}n^\lambda)(h_{\nu\rho}n^\rho) + h_{\mu\nu}. \tag{B.4}$$

It is easy to check that $g_{\mu\nu}$ is then a Lorentzian metric, and that n^μ is a unit timelike vector.

A time orientation of M amounts to a continuous choice of the sign of n. Thus M will admit a time-orientable Lorentzian metric if and only if it admits a nowhere vanishing vector field. This makes it possible to use the Poincaré–Hopf index theorem (see appendix A) to determine which topologies are potential candidates for $(2+1)$-dimensional spacetimes.

Consider first a noncompact manifold M. A typical vector field on M will have zeros, but these zeros can always be 'pushed off to infinity'. Given a vector field n with an isolated zero at some point x, we can find a sequence of points $\{x_i\}$ starting at x and having no limit point, and a sequence of diffeomorphisms that successively move the zero from x_i to x_{i+1}. Taking the limit, and repeating for each zero of n, we obtain a vector field on M with no zeros, and hence a Lorentzian metric.

If M is closed – that is, compact and without boundary – then the Poincaré–Hopf index theorem tells us that a nowhere vanishing vector field exists if and only if the Euler number $\chi(M)$ vanishes. But the Euler number of an odd-dimensional closed manifold is always zero, so we can again obtain a nowhere vanishing vector field, and hence a time-orientable Lorentzian metric.

If M is compact and has a boundary, on the other hand, both the mathematics and the physics become a bit more complicated. The natural physical question is now this: given a three-manifold M with a boundary $\partial M = \Sigma^- \cup \Sigma^+$, can one find a time-orientable Lorentzian metric such that Σ^- is the spacelike past boundary of M and Σ^+ is the spacelike future boundary? To answer this question we can use the same kinds of topological tools, but we will now have to pay attention to the direction of the vector field n on ∂M, since this field determines the direction of time and thus distinguishes past and future boundaries.

To analyze this case, we need a slight generalization of the Poincaré–Hopf index theorem [242]. Recall from appendix A that if v is a vector field pointing outward at each boundary component of M, the index of v – the sum of the degree of $v/|v|$ at each of its zeros – satisfies

$$ind\,(v) = \chi(M), \tag{B.5}$$

and v can be deformed to a nowhere vanishing vector field if and only if $ind\,(v) = 0$. To obtain a corresponding result for a vector field n pointing *inward* at Σ^-, let us first extend M slightly at this boundary. That is, if $N = [0,\epsilon] \times \Sigma^-$ is the three-dimensional 'cylinder' with cross-section Σ^-, we construct a new manifold

$$\widehat{M} = M \cup_{\Sigma^-} N \tag{B.6}$$

by attaching N to M along Σ^-. It is clear that any vector field n on M that points inward at Σ^- can be extended to a vector field $\hat{n}$ on $\widehat{M}$, perhaps with new zeros in N, that points outward at the new boundary. Then since both $\hat{n}$ and its restriction $\hat{n}|_N$ to N point outward at the relevant boundaries, we can apply the Poincaré–Hopf theorem to $\widehat{M}$ and N, obtaining

$$ind\,(\hat{n}) = \chi(\widehat{M})$$
$$ind\,(\hat{n}|_N) = \chi(N). \tag{B.7}$$

But it is clear from the definition of the index that $ind\,(\hat{n}) = ind\,(n) + ind\,(\hat{n}|_N)$, so

$$ind\,(n) = \chi(\widehat{M}) - \chi(N) = \chi(M) - \chi(N), \tag{B.8}$$

where the last equality comes from the fact the $\widehat{M}$ is homeomorphic to M. Finally, we use the fact that $\chi(N) = \chi(\Sigma^-)$ (the Cartesian product with the interval 'adds no topology') to obtain the relationship

$$ind\,(n) = \chi(M) - \chi(\Sigma^-). \tag{B.9}$$

Reversing the direction of n, the same argument shows that

$$ind\,(-n) = \chi(M) - \chi(\Sigma^+). \tag{B.10}$$

But it is also the case that in odd dimensions, $ind\,(-n) = -ind\,(n)$, so combining (B.9) and (B.10), we see that

$$ind\,(n) = \frac{1}{2}\left(\chi(\Sigma^+) - \chi(\Sigma^-)\right). \tag{B.11}$$

For n to have no zeros, this index must vanish. We have thus shown that a compact three-manifold M with boundary $\Sigma^- \cup \Sigma^+$ admits a time-orientable Lorentzian metric with past spacelike boundary Σ^- and future spacelike boundary Σ^+ if and only if $\chi(\Sigma^+) = \chi(\Sigma^-)$. In short, time-orientable Lorentzian metrics on compact spacetimes require 'conservation of Euler number'.

B.2 Lorentz cobordism

This description of spatial topology change suggests a related, but slightly different, question. In the preceding section, we started with a three-manifold M with a given boundary $\Sigma^- \cup \Sigma^+$, and asked whether a suitable Lorentzian metric existed on M. We could instead start with two surfaces Σ^- and Σ^+ and ask whether *any* Lorentzian manifold interpolates between them [242, 296].

Let M_1 and M_2 be two (not necessarily connected) n-manifolds. A *cobordism* between M_1 and M_2 is an $(n + 1)$-dimensional manifold M whose boundary is the disjoint union $\partial M = M_1 \cup M_2$. A *Lorentz cobordism* is a cobordism M along with a Lorentzian metric on M such that M_1 and M_2 are spacelike, M_1 is the past boundary of M, and M_2 is the future boundary.

In four or more dimensions, not every pair of manifolds is cobordant. For $n < 4$, however, it can be shown that any two orientable manifolds are cobordant [197]. In particular, there exists a cobordism between any pair of orientable surfaces. Such a cobordism is easy to visualize: to connect a surface of genus g_1 to one of genus g_2, simply take a genus g_1 handlebody (that is, a 'solid genus g_1 surface') and cut out a genus g_2 handlebody from its interior.

Combining this fact with the results of the previous section, we see that a compact time-oriented Lorentz cobordism exists between two surfaces Σ^- and Σ^+ if and only if $\chi(\Sigma^-) = \chi(\Sigma^+)$. This cobordism is by no means unique: given one manifold M connecting Σ^- and Σ^+, we can change the topology of M away from the boundaries – for instance, by taking a connected sum with a closed orientable three-manifold – and the results of the last section guarantee that we can again find a Lorentzian metric on the new manifold. There are thus a huge (in fact, infinite) number of Lorentzian manifolds joining initial and final surfaces with the same Euler number.

A *noncompact* Lorentz cobordism, moreover, can be found between *any* two orientable surfaces, even if their Euler numbers differ. Let us take any compact manifold M that is a cobordism between Σ^- and Σ^+ and try to construct a Lorentzian metric on M. Proceeding as in the last section, we will obtain a vector field n pointing inward at Σ^- and outward at Σ^+, but now with a nonzero index; that is to say, n will now have zeros at some set of points $x_i \in M$. But we can 'kill' these zeros by simply cutting out the points x_i, giving us a new noncompact manifold $M' = M \backslash \{x_i\}$ with a Lorentzian metric. Of course, the missing points make M' rather physically uninteresting, but we can push these points off to infinity by means of a conformal transformation to obtain a somewhat more reasonable spacetime. Alternatively, Sorkin has pointed out that if we drop the requirement of time-orientability, the zeros of n can be eliminated by cutting out a small ball around each point x_i and identifying the antipodal points of the resulting spherical boundary. This process will give us a new compact manifold M'' that is topologically the connected sum of M with copies of real projective space $\mathbf{R}P^3$. The vector field n will not be well-behaved under such identifications, of course, but the line element field $(n, -n)$ will be, so we will be able to find a non-time-orientable Lorentzian metric on M''.

B.3 Closed timelike curves and causal structure

The topology changes we have found come at a cost, however: topology-changing universes typically involve causality violations. We have seen that if $\chi(\Sigma^+) \neq \chi(\Sigma^-)$, a topology-changing spacetime must either have some 'missing points' or else be non-time-orientable. Geroch has found an even stronger result: if compact initial and final surfaces Σ^- and Σ^+ are not homeomorphic, then any compact time-oriented Lorentz cobordism between them must contain closed timelike curves [122].

The proof of this result is reasonably straightforward. We have seen that a time-oriented Lorentzian metric on M determines a timelike vector field n. If M is compact and contains no closed timelike curves, an integral curve of n starting at a point $x \in \Sigma^-$ must end at some point $x' \in \Sigma^+$. The mapping $x \mapsto x'$ is then the required homeomorphism between Σ^- and Σ^+. Moreover, by following the integral curves forward from Σ^-, we can obtain a complete time-slicing of M, thus proving that M must have the product topology $M \approx [0, 1] \times \Sigma$.

Closed timelike curves signal an obvious violation of causality. To further describe the causal structure of spacetime, it is useful to define several new mathematical objects. (For more detail, see [123] or [145]; note, though, that some definitions differ slightly from source to source.) In particular, we need some notion of 'space at a given time' in order to ask whether the future can be completely determined from data 'now'.

A *slice* S of an n-dimensional spacetime M is an embedded spacelike $(n-1)$-dimensional submanifold that is closed as a subset of M. The condition of closure is required to eliminate 'small' submanifolds such as the disk $x^2 + y^2 < 1$ in $\mathbf{R}^3$. (The closed disk $x^2 + y^2 \leq 1$ is excluded

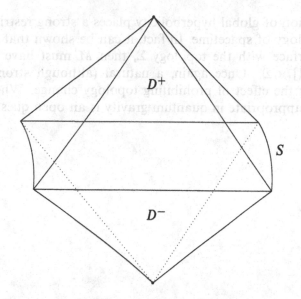

Fig. B.1. The domain of dependence of the slice S is the region in which all physics can be predicted or retrodicted from data on S.

because it is not a manifold, since it has a boundary.) The *future domain of dependence* of a slice S, $D^+(S)$, is the set of points p such that every past-directed timelike curve from p with no past endpoint* meets S (see figure B.1). Physically, this means that events in $D^+(S)$ are fully determined by conditions on S – any signal that can reach a point p in $D^+(S)$ must have previously passed through S. The *past domain of dependence* $D^-(S)$, is defined similarly; the *domain of dependence* of S is

$$D(S) = D^+(S) \cup D^-(S). \qquad (B.12)$$

Points in $D(S)$ are thus those at which all physics can be predicted or retrodicted from a complete knowledge of physics on S.

A *Cauchy surface* is a slice whose domain of dependence is all of spacetime. Not every spacetime admits a Cauchy surface; one that does is *globally hyperbolic*. Physically, the information on a Cauchy surface should suffice to determine all events throughout M; for instance, the specification of initial data for the Maxwell equations on a Cauchy surface fixes the value of the electromagnetic field throughout spacetime. The absence of a Cauchy surface signals a breakdown in global predictability, in the sense that complete data at any fixed time does not determine the entire past or future.

---

* A curve with a past endpoint is one that 'stops' not because of any feature of M, but merely because we have not defined it to go further.

The condition of global hyperbolicity places a strong restriction on the possible topology of spacetime. In fact, it can be shown that if M admits a Cauchy surface with the topology Σ, then M must have the product topology $[0, 1] \times \Sigma$. Once again, a natural (although strong) causality condition has the effect of prohibiting topology change. Whether such a restriction is appropriate in quantum gravity is an open question.

Appendix C
Differential geometry and fiber bundles

The differential geometry required for general relativity is discussed in most textbooks – see, for instance, [198] or [274] – and there are a number of other good references available [76, 105, 159]. I will not attempt to repeat that material here. The purpose of this appendix is rather to establish notation and conventions and to stress a few aspects of differential geometry that are not normally covered in such introductions.

C.1 Parallel transport and connections

Let M be an n-dimensional differentiable manifold, upon which the usual structure of vectors, tensors, and differential forms has been defined. In order to describe the geometry of M, we must be able to compare vectors at different points; that is, we must have a procedure for transporting a vector from one point to another. A vector v^μ is said to be parallel transported along a curve $x^\mu = x^\mu(s)$ if it satisfies the equation

$$\frac{dv^\rho}{ds} + \Gamma^\rho_{\mu\nu} \frac{dx^\mu}{ds} v^\nu = 0, \qquad (C.1)$$

where the $\Gamma^\rho_{\mu\nu}$ is the *connection*.* Expression (C.1) can be converted to an integral equation and solved by iteration,

$$v^\rho(s) = v^\rho(0) - \int_0^s ds_1 \Gamma^\rho_{\mu\nu}(x(s_1)) \frac{dx^\mu}{ds_1} v^\nu(s_1)$$

$$= v^\rho(0) - \int_0^s ds_1 \Gamma^\rho_{\mu\nu}(x(s_1)) \frac{dx^\mu}{ds_1} \left(v^\nu(0) - \int_0^{s_1} ds_2 \Gamma^\nu_{\sigma\tau}(x(s_2)) \frac{dx^\sigma}{ds_2} v^\tau(s_2) \right)$$

$$= \cdots = U^\rho{}_\nu(s,0) v^\nu(0), \quad (C.2)$$

---

* Here, as elsewhere in this book, Greek letters $\mu, \nu, \rho, \ldots$ denote indices in a coordinate basis. Many of the expressions below are somewhat more complicated in a noncoordinate basis, but the more general results can be found in standard references.

243

where

$$U^\rho{}_\nu(s_1, s_2) = P \exp\left\{-\int_{s_2}^{s_1} \Gamma^\rho_{\mu\nu} \frac{dx^\mu}{ds} ds\right\}. \tag{C.3}$$

Here the symbol P denotes path ordering,

$$P(A(s_1)B(s_2)) = \begin{cases} A(s_1)B(s_2) & \text{if } s_1 > s_2 \\ B(s_2)A(s_1) & \text{if } s_2 > s_1 \end{cases}, \tag{C.4}$$

and the path-ordered exponent is defined by its Taylor expansion,

$$P \exp\left\{\int_0^s ds_1 A_\mu \frac{dx^\mu}{ds_1}\right\}$$
$$= \sum_{n=0}^\infty \frac{1}{n!} \int_0^s ds_1 \cdots \int_0^s ds_n \, P\left(A_{\mu_1} \frac{dx^{\mu_1}}{ds_1} \cdots A_{\mu_n} \frac{dx^{\mu_n}}{ds_n}\right). \tag{C.5}$$

The parallel transport matrix U is rarely discussed in introductory texts on differential geometry, but the derivation (C.2) is standard in quantum mechanics, where it is used to obtain the scattering matrix; see, for example, section 3.5 of reference [278].

Given a concept of parallel transport and a choice of connection, we can now define the *covariant derivative*. The covariant derivative of a vector field v^μ in the direction tangent to a curve $x^\mu = x^\mu(s)$, for example, is

$$\frac{dx^\mu}{ds} \nabla_\mu v^\rho = \lim_{\epsilon \to 0} \frac{1}{\epsilon} [v^\rho(x(s+\epsilon)) - U^\rho{}_\nu(s+\epsilon, s) v^\nu(s)]$$
$$= \frac{dx^\mu}{ds}\left(\partial_\mu v^\rho + \Gamma^\rho_{\mu\nu} v^\nu\right). \tag{C.6}$$

Note that the parallel transport matrix is needed for such a definition to make sense: otherwise, the expression in brackets in (C.6) would involve a difference between vectors at two different points, which is not defined unless one has a canonical procedure for comparing different tangent spaces. Covariant derivatives of other types of tensors can be determined by the Leibnitz rule and the rule that the covariant derivative of a function f is just its ordinary derivative.

The *torsion* $T^\rho_{\mu\nu}$ of a connection is defined by the equation

$$(\nabla_\mu \nabla_\nu - \nabla_\nu \nabla_\mu)f = T^\rho_{\mu\nu} \nabla_\rho f \tag{C.7}$$

for any function f (again, in a coordinate basis). If the torsion vanishes, $\Gamma^\rho_{\mu\nu}$ is symmetric in its lower two indices. Given a metric $g_{\mu\nu}$, the condition

Differential geometry and fiber bundles 245

of compatibility with the connection,

$$\nabla_\rho g_{\mu\nu} = 0,$$ (C.8)

determines a unique torsion-free connection, the Christoffel connection.

A *geodesic* on M is a curve whose tangent vector is parallel transported, that is,

$$u^\rho \nabla_\rho u^\mu = 0, \qquad u^\mu = \frac{dx^\mu}{ds}.$$ (C.9)

When the connection is written explicitly, this gives equation (1.9). If the connection is compatible with a metric $g_{\mu\nu}$, a geodesic is also a curve of extremal arc length.

C.2 Holonomy and curvature

If $\gamma : [0,1] \to M$ is a closed curve, the parallel transport matrix $U(1,0)$ is known as the *holonomy matrix* of γ. (This is one of several different meanings of the word 'holonomy'.) If M is flat Euclidean or Minkowski space, the holonomy of every curve is the identity. In general, the holonomies of the set of loops based at a point p form a group, the holonomy group.

Suppose γ is an infinitesimal closed curve that encloses a surface S in M. The holonomy of γ can be computed to lowest order by expanding the exponential (C.3) and using Stokes' theorem, yielding

$$U^\rho{}_\nu \approx \delta^\rho_\nu - \int_S R^\rho{}_{\nu\sigma\tau} dS^{\sigma\tau},$$ (C.10)

where

$$R^\rho{}_{\nu\sigma\tau} = \partial_\sigma \Gamma^\rho_{\tau\nu} - \partial_\tau \Gamma^\rho_{\sigma\nu} - \Gamma^\mu_{\sigma\nu}\Gamma^\rho_{\tau\mu} + \Gamma^\mu_{\tau\nu}\Gamma^\rho_{\sigma\mu}$$ (C.11)

is the *Riemann curvature tensor*. The Ricci tensor is the contraction

$$R_{\mu\nu} = R^\rho{}_{\mu\rho\nu},$$ (C.12)

and the curvature scalar is

$$R = g^{\mu\nu} R_{\mu\nu}.$$ (C.13)

The curvature tensor also determines commutators of covariant derivatives: for instance,

$$[\nabla_\rho, \nabla_\mu]v^\sigma = R^\sigma{}_{\nu\rho\mu}v^\nu.$$ (C.14)

If a manifold M is flat – that is, if the curvature tensor $R^\rho{}_{\nu\sigma\tau}$ vanishes everywhere – and M is also topologically trivial, then M is necessarily a region of Euclidean (or Minkowski) space. If $\pi_1(M) \neq 0$, on the other hand, the vanishing of the curvature does not uniquely determine the geometry: a noncontractible curve γ does not enclose a surface, and we cannot use Stokes' theorem to evaluate its holonomy.

C.3 Frames and spin connections

It is often useful to introduce a local *frame*, that is, a set of orthonormal vectors $e^{\mu a}$ that satisfy

$$g^{\mu\nu} e_\mu{}^a e_\nu{}^b = \eta^{ab}, \quad e_\mu{}^a e_\nu{}^b \eta_{ab} = g_{\mu\nu}. \tag{C.15}$$

In three dimensions, such a frame is often called a triad or dreibein. In general, the $e_\mu{}^a$ cannot have vanishing covariant derivatives, since this would imply flatness. Rather, we can write

$$\nabla_\mu e_\nu{}^a + \omega_\mu{}^a{}_b e_\nu{}^b = 0, \tag{C.16}$$

where $\omega_\mu{}^a{}_b$ is the *spin connection*. If the connection $\Gamma^\rho_{\mu\nu}$ is torsion-free, it is easy to see that

$$D_\omega e^a = de^a + \omega^a{}_b \wedge e^b = 0, \tag{C.17}$$

where $e^a = e_\mu{}^a dx^\mu$ is the frame one-form and D_ω, defined by (C.17), is the gauge-covariant exterior derivative. If $\Gamma^\rho_{\mu\nu}$ is not torsion-free, it may be checked that $D_\omega e^a$ is proportional to the torsion. Equation (C.16) can be solved for $\omega_\mu{}^a{}_b$ in terms of the metric and the torsion; in 2+1 dimensions, the solution for the torsion-free case is given by equation (2.63).

The curvature may be obtained in terms of the spin connection by computing the commutator $[\nabla_\mu, \nabla_\nu] e_\rho{}^a$. A simple calculation gives

$$R^a{}_b = e_\mu{}^a e_{vb} R^{\mu\nu}{}_{\rho\sigma} dx^\rho \wedge dx^\sigma = d\omega^a{}_b + \omega^a{}_c \wedge \omega^c{}_b, \tag{C.18}$$

where $\omega^a{}_b = \omega_\mu{}^a{}_b dx^\mu$ is the spin connection one-form. Equations (C.17) and (C.18) are the Cartan structure equations.

C.4 The tangent bundle

Many elements of differential geometry and gauge theory are best expressed in the language of fiber bundles. The archetypical fiber bundle is the *tangent bundle* TM of an n-manifold M. A point in TM consists of a pair (p, v_p), where p is a point in M and v_p is a tangent vector at p. There is a natural projection $\pi : TM \to M$ that takes (p, v_p) to p. The inverse image $\pi^{-1}(p)$ is known as the fiber at p; for the tangent bundle, it is just the tangent space T_pM. Locally, TM looks like $U \times \mathbf{R}^n$, where U is an open set in M. This product structure generally cannot be extended globally, however; that is, TM as a whole is typically not a product manifold $M \times \mathbf{R}^n$.

The *cotangent bundle* T^*M has a similar definition; its points are pairs (p, ω_p), where ω_p is a covariant vector (or one-form) at p. Similarly, the *frame bundle* consists of pairs (p, e_p^a), where e_p^a is a frame at p.

Much of the differential geometry summarized in the preceding sections has a natural description in terms of TM. In particular, parallel transport can be understood as a procedure for lifting curves from M to TM. Consider a curve γ in M and a tangent vector v at $\gamma(0)$. By definition, the pair $(\gamma(0), v)$ is a point in TM. If we now parallel transport v along γ, we obtain a new vector $U(s,0)v$ at each point $\gamma(s)$, and thus a curve $(\gamma(s), U(s,0)v)$ in TM. In contrast to equation (C.1), this 'lifting' description does not require an explicit choice of coordinates, and is thus useful for analyzing global issues.

C.5 Fiber bundles and connections

A *fiber bundle* is a manifold E with a local product structure that generalizes the properties of the tangent bundle. More precisely, a fiber bundle with total space E, base space M, and fiber F consists of a space E and a projection $\pi : E \to M$ that is locally trivial – for each point $x \in M$, there is a neighborhood $U \subset M$ and a homeomorphism $\Phi : U \times F \to \pi^{-1}(U)$ such that

$$\pi \circ \Phi(x,y) = x, \qquad x \in U, \; y \in F. \tag{C.19}$$

In other words, the region $\pi^{-1}(U)$ looks like a product $U \times F$, with each fiber $\pi^{-1}(x)$ homeomorphic to F, but this product description need not be globally valid.

Given two overlapping neighborhoods U_i and U_j in M, one can form the transition function

$$\Phi_{ij} = \Phi_i^{-1} \circ \Phi_j : (U_i \cap U_j) \times F \to (U_i \cap U_j) \times F. \tag{C.20}$$

For each fixed $x \in U_i \cap U_j$, Φ_{ij} maps F to itself. We require that the transition functions be elements of some group G of transformations of the fiber F; G is called the *structure group*. If F is a vector space and G is a group of linear maps, then E is called a *vector bundle*. If F is itself a group G, and the transition functions $\Phi \in G$ act by left multiplication, then E is a *principle bundle*. The tangent bundle TM is a vector bundle; the frame bundle can be viewed as a principle $GL(n, \mathbf{R})$ bundle.

A *section* (or cross-section) of a bundle is a map $\sigma : M \rightarrow E$ such that the image of each point $x \in M$ lies in the fiber $\pi^{-1}(x)$ over x, that is,

$$\pi \circ \sigma(x) = x, \qquad x \in M. \tag{C.21}$$

Vector fields are thus sections of the tangent bundle, while charged fields in quantum electrodynamics are sections of $U(1)$ bundles.

There are a number of equivalent definitions of a connection on a fiber bundle, but for the purposes of this book a fairly simple approach is sufficient. Let E be a vector bundle, and let U be a neighborhood in M such that $\pi^{-1}(U) \approx U \times F$. We can choose a trivialization Φ and use it to identify $\pi^{-1}(U)$ with $U \times F$; this is the analog of choosing a particular coordinate system in elementary differential geometry. The fiber F is a vector space, and carries a representation of the structure group G. If we choose a basis e_i of F, points in $\pi^{-1}(U)$ can be represented as pairs (x^μ, v^i), and the generators of G will be matrices $(T_a)^i{}_j$.

To define parallel transport in E, we again let $x^\mu = x^\mu(s)$ denote a curve in M, and choose an initial vector v^i at $x^\mu(0)$. The parallel transport equation analogous to (C.1) is then

$$\frac{dv^i}{ds} + \frac{dx^\mu}{ds} A_\mu{}^a T_a{}^i{}_j v^j = 0, \tag{C.22}$$

where $A_\mu{}^a$ is again called a connection. Once again, parallel transport defines a lifting of the curve $x^\mu(s)$ from M to E. As in section 1, equation (C.22) can be solved in terms of a path-ordered exponent,

$$v(s) = U(s,0)v(0)$$
$$U(s_1, s_2) = P \exp\left\{ -\int_{s_1}^{s_2} A_\mu{}^a T_a \frac{dx^\mu}{ds} ds \right\}, \tag{C.23}$$

where I have suppressed the fiber indices. The concepts of covariant derivative, holonomy, and curvature all generalize in a straightforward manner. In particular, if

$$D_\mu v = \partial_\mu v + A_\mu{}^a T_a v \tag{C.24}$$

is the covariant derivative, it is easy to check that

$$[D_\mu, D_\nu]v = \left(\partial_\mu A_\nu{}^a - \partial_\nu A_\mu{}^a + c_{bc}{}^a A_\mu{}^b A_\nu{}^c \right) T_a v = F_{\mu\nu}{}^a T_a v, \tag{C.25}$$

where $c_{bc}{}^a$ are the structure constants of the group G. Equation (C.25) shows that the familiar gauge field strength tensor can be interpreted as the curvature of a bundle connection. Moreover, if $\gamma : [0,1] \rightarrow M$

is a closed curve, the parallel transport matrix $U(1,0)$ again gives the holonomy, and the exponential (C.23) may be recognized as the 'Wilson loop' (4.60). The idea that a holonomy may be nonzero even when $F_{\mu\nu}$ vanishes is familiar in gauge theories: the holonomy is then precisely the Aharonov–Bohm phase shift [6].

References

[1] Achúcarro, A. and Townsend, P. K. (1986). A Chern–Simons action for three-dimensional anti-de Sitter supergravity theories, *Phys. Lett.* **B180**, 89–92.

[2] Alvarez, O. (1986). Differential geometry in string models, in *Unified string theories*, ed. M. Green and D. Gross (World Scientific, Singapore).

[3] Amano, A. and Higuchi, S. (1992a). ISO(2, 1) gauge fields and (2+1)-dimensional space-time, *Prog. Theor. Phys. Suppl.* **110**, 151–157.

[4] Amano, A. and Higuchi, S. (1992b). Topology change in ISO(2, 1) Chern–Simons gravity, *Nucl. Phys.* **B377**, 218–235.

[5] Ambjørn, J., Jurkiewicz, J., and Watabiki, Y. (1995). Dynamical triangulations, a gateway to quantum gravity?, *J. Math. Phys.* **36**, 6299–6339.

[6] Anandan, J. S. (1993). Remarks concerning the geometries of gravity and gauge fields, in *Directions in general relativity*, ed. B. L. Hu, M. P. Ryan Jr., and C. V. Vishveshwara (Cambridge University Press, Cambridge).

[7] Anderson, A. (1993). Unitary equivalence of the metric and holonomy formulations of (2+1)-dimensional quantum gravity on the torus, *Phys. Rev.* **D47**, 4458–4470.

[8] Andersson, L., Moncrief, V., and Tromba, A. J. (1997). On the global evolution problem in 2+1 gravity, *J. Geom. Phys.* **23**, 191–205.

[9] Arnowitt, R., Deser, S., and Misner, C. W. (1962). The dynamics of general relativity, in *Gravitation: an introduction to current research*, ed. L. Witten (Wiley, New York).

[10] Ashtekar, A. (1990). Lessons from (2+1)-dimensional quantum gravity, in *Strings '90*, ed. R. Arnowitt, R. Bryan, M. J. Duff *et al.* (World Scientific, Singapore).

[11] Ashtekar, A. (1991). *Lectures on non-perturbative canonical gravity* (World Scientific, Singapore).

250

[12] Ashtekar, A. (1996). Large quantum gravity effects: unexpected limitations of the classical theory, *Phys. Rev. Lett.* **77**, 4864–4867.

[13] Ashtekar, A., Husain, V., Rovelli, C., Samuel, J., and Smolin, L. (1989). (2+1) quantum gravity as a toy model for the (3+1) theory, *Class. Quant. Grav.* **6**, L185–L193.

[14] Ashtekar, A. and Loll, R. (1994). New loop representations for (2+1) gravity, *Class. Quant. Grav.* **11**, 2417–2434.

[15] Ashtekar, A. and Magnon, A. (1975). Quantum fields in curved space-times, *Proc. Roy. Soc. Lond.* **A346**, 375–394.

[16] Ashtekar, A. and Pierri, M. (1996). Probing quantum gravity through exactly soluble midisuperspaces, *J. Math. Phys.* **37**, 6250–6270.

[17] Atiyah, M. (1990). *The geometry and physics of knots* (Cambridge University Press, Cambridge).

[18] Auslander, L. and Markus, L. (1955). Holonomy of flat affinely connected manifolds, *Am. J. Math.* **62**, 139–159.

[19] Avis, S. J., Isham, C. J., and Storey, D. (1978). Quantum field theory in anti-de Sitter space-time, *Phys. Rev.* **D18**, 3565–3576.

[20] Balachandran, A. P., Chandar, L., and Momen, A. (1996). Edge states in gravity and black hole physics, *Nucl. Phys.* **B461**, 581–596.

[21] Bañados, M. and Gomberoff, A. (1997). Black hole entropy in the Chern–Simons formulation of (2+1)-dimensional gravity, *Phys. Rev.* **D55**, 6162–6167.

[22] Bañados, M., Henneaux, M., Teitelboim, C., and Zanelli, J. (1993). Geometry of the (2+1) black hole, *Phys. Rev.* **D48**, 1506–1525.

[23] Bañados, M., Teitelboim, C., and Zanelli, J. (1992). The black hole in three-dimensional space-time, *Phys. Rev. Lett.* **69**, 1849–1851.

[24] Banks, T., Fischler, W., and Susskind, L. (1985). Quantum cosmology in 2+1 and 3+1 dimensions, *Nucl. Phys.* **B262**, 159–186.

[25] Barbero, J. F. and Varadarajan, M. (1994). The phase space of (2+1)-dimensional gravity in the Ashtekar formulation, *Nucl. Phys.* **B415**, 515–532.

[26] Barbero, J. F. and Varadarajan, M. (1995). Homogeneous (2+1)-dimensional gravity in the Ashtekar formulation, *Nucl. Phys.* **B456**, 355–376.

[27] Barrett, J. W. (1995). Quantum gravity as a topological quantum field theory, *J. Math. Phys.* **36**, 6161–6179.

[28] Barrett, J. W. and Crane, L. (1997). An algebraic interpretation of the Wheeler–DeWitt equation, *Class. Quant. Grav.* **14**, 2113–2121.

[29] Barrett, J. W. and Foxon, T. J. (1994). Semiclassical limits of simplicial quantum gravity, *Class. Quant. Grav.* **11**, 543–556.

[30] Barrow, J. D., Burd, A. B., and Lancaster, D. (1986). Three-dimensional classical spacetimes, *Class. Quant. Grav.* **3**, 551–567.

[31] Becker, K., Becker, M., and Strominger, A. (1995). Three-dimensional supergravity and the cosmological constant, *Phys. Rev.* **D51**, R6603–R6607.

[32] Bekenstein, J. D. (1973). Black holes and entropy, *Phys. Rev.* **D7**, 2333–2346.

[33] Bellini, A., Ciafaloni, M., and Valtancoli, P. (1995). (2+1) gravity with moving particles in an instantaneous gauge, *Nucl. Phys.* **B454**, 449–464.

[34] Bellini, A., Ciafaloni, M., and Valtancoli, P. (1996). Solving the N-body problem in (2+1) gravity, *Nucl. Phys.* **B462**, 453–492.

[35] Bengtsson, I. (1989). Yang–Mills theory and general relativity in three and four dimensions, *Phys. Lett.* **B220**, 51–54.

[36] Berger, M. and Ebin, D. (1969). Some decompositions of the space of symmetric tensors on a Riemannian manifold, *J. Diff. Geom.* **3**, 379–392.

[37] Birman, J. S. (1977). The algebraic structure of surface mapping class groups, in *Discrete groups and automorphic functions*, ed. W. J. Harvey (Academic Press, London).

[38] Birmingham, D., Cho, H. T., and Kantowski, R. (1991). The Vilkovisky–DeWitt effective action in BF-type topological field theories, *Phys. Lett.* **B269**, 116–122.

[39] Birmingham, D., Gibbs, R., and Mokhtari, S. (1991). The effective action in (2+1)-dimensional gravity and generalized BF topological field theory, *Phys. Lett.* **B263**, 176–182.

[40] Brans, C. H. (1994). Exotic smoothness and physics, *J. Math. Phys.* **35**, 5494–5506.

[41] Brown, J. D. (1988). *Lower dimensional gravity* (World Scientific, Singapore).

[42] Brown, J. D. and Henneaux, M. (1986). Central charges in the canonical realization of asymptotic symmetries: an example from three dimensional gravity, *Commun. Math. Phys.* **104**, 207–226.

[43] Brown, J. D. and York, J. W. (1993a). Quasilocal energy and conserved charges derived from the gravitational action, *Phys. Rev.* **D47**, 1407–1419.

[44] Brown, J. D. and York, J. W. (1993b). Microcanonical functional integral for the gravitational field, *Phys. Rev.* **D47**, 1420–1431.

[45] Boulatov, D. V. (1992). A model of three-dimensional lattice gravity, *Mod. Phys. Lett.* **A7**, 1629–1646.

[46] Buchbinder, I. L., Lyakhovich, S. L., and Krykhtin, V. A. (1993). Canonical quantization of topologically massive gravity, *Class. Quant. Grav.* **10**, 2083–2090.

[47] Cangemi, D., Leblanc, M., and Mann, R. B. (1993). Gauge formulation of the spinning black hole in (2+1)-dimensional anti-de Sitter space, *Phys. Rev.* **D48**, 3606–3610.

[48] Carfora, M. and Marzuoli, A. (1993). Critical phenomena for Riemanian manifolds: simple homotopy and simplicial quantum gravity, preprint gr-qc/9303029.

[49] Carfora, M. and Marzuoli, A. (1995a). Entropy estimates for simplicial quantum gravity, *J. Geom. Phys.* **16**, 99–119.

[50] Carfora, M. and Marzuoli, A. (1995b). Holonomy and entropy estimates for dynamically triangulated manifolds, *J. Math. Phys.* **36**, 6353–6376.

[51] Carlip, S. (1989). Exact quantum scattering in 2+1 dimensional gravity, *Nucl. Phys.* **B324**, 106–122.

[52] Carlip, S. (1990). Observables, gauge invariance, and time in (2+1)-dimensional gravity, *Phys. Rev.* **D42**, 2647–2654.

[53] Carlip, S. (1992a). (2+1)-dimensional Chern–Simons gravity as a Dirac square root, *Phys. Rev.* **D45**, 3584–3590; erratum *Phys. Rev.* **D47**, 1729.

[54] Carlip, S. (1992b). Entropy versus action in the (2+1)-dimensional Hartle–Hawking wave function, *Phys. Rev.* **D46**, 4387–4395.

[55] Carlip, S. (1993a). Modular group, operator ordering, and time in (2+1)-dimensional gravity, *Phys. Rev.* **D47**, 4520–4524.

[56] Carlip, S. (1993b). The sum over topologies in three-dimensional Euclidean quantum gravity, *Class. Quant. Grav.* **10**, 207–218.

[57] Carlip, S. (1993c). Real tunneling solutions and the Hartle–Hawking wave function, *Class. Quant. Grav.* **10**, 1057–1064.

[58] Carlip, S. (1994a). Geometric structures and loop variables in (2+1)-dimensional gravity, in *Knots and quantum gravity*, ed. J. Baez (Oxford University Press, Oxford).

[59] Carlip, S. (1994b). Notes on the (2+1)-dimensional Wheeler–DeWitt equation, *Class. Quant. Grav.* **11**, 31–40.

[60] Carlip, S. (1995a). Statistical mechanics of the (2+1)-dimensional black hole, *Phys. Rev.* **D51**, 632–637.

[61] Carlip, S. (1995b). A phase space path integral for (2+1)-dimensional gravity, *Class. Quant. Grav.* **12**, 2201–2207.

[62] Carlip, S. (1995c). The (2+1)-dimensional black hole, *Class. Quant. Grav.* **12**, 2853–2879.

[63] Carlip, S. (1997). The statistical mechanics of the three-dimensional Euclidean black hole, *Phys. Rev.* **D55**, 878–882.

[64] Carlip, S. (1997). Spacetime foam and the cosmological constant, *Phys. Rev. Lett.* **79**, 4071–4074.

[65] Carlip, S., Clements, M., DellaPietra, S., and DellaPietra, V. (1990). Sewing Polyakov amplitudes I: Sewing at a fixed conformal structure, *Commun. Math. Phys.* **127**, 253–271.

[66] Carlip, S. and Cosgrove, R. (1994). Topology change in (2+1)-dimensional gravity, *J. Math. Phys.* **35**, 5477–5493.

[67] Carlip, S. and de Alwis, S. P. (1990). Wormholes in 2+1 dimensions, *Nucl. Phys.* **B337**, 681–694.

[68] Carlip, S. and Nelson, J. E. (1994). Equivalent quantisations of (2+1)-dimensional gravity, *Phys. Lett.* **B324**, 299–302.

[69] Carlip, S. and Nelson, J. E. (1995). Comparative quantizations of (2+1)-dimensional gravity, *Phys. Rev.* **D51**, 5643–5653.

[70] Carlip, S. and Nelson, J. E. (1997). The quantum modular group in (2+1)-dimensional gravity, in preparation.

[71] Carlip, S. and Teitelboim, C. (1995). Aspects of black hole quantum mechanics and thermodynamics in 2+1 dimensions, *Phys. Rev.* **D51**, 622–631.

[72] Carter, J. S., Flath, D. E., and Saito, M. (1995). *The classical and quantum 6j-symbols*, Mathematical Notes 43 (Princeton University Press, Princeton).

[73] Catterall, S. (1995). Simulations of dynamically triangulated gravity – an algorithm for arbitrary dimension, *Comput. Phys. Comm.* **87**, 409–415.

[74] Chan, K. C. K. (1997). Modifications of the BTZ black hole by a dilaton and scalar, *Phys. Rev.* **D55**, 3564–3574.

[75] Chan, K. C. K. and Mann, R. B. (1996). Spinning black holes in (2+1)-dimensional string and dilaton gravity, *Phys. Lett.* **B371**, 199–205.

[76] Choquet-Bruhat, Y., DeWitt-Morette, C., and Dillard-Bleick, M. (1977). *Analysis, manifolds and physics* (North-Holland, Amsterdam).

[77] Clement, G. (1976). Field-theoretic extended particles in two space dimensions, *Nucl. Phys.* **B114**, 437–448.

[78] Clement, G. (1985). Stationary solutions in three-dimensional general relativity, *Int. J. Theor. Phys.* **24**, 267–275.

[79] Collas, P. (1977). General relativity in two- and three-dimensional spacetimes, *Am. J. Phys.* **45**, 833–837.

[80] Cornfeld, I. P., Fomin, S. V., and Sinai, Ya. G. (1982). *Ergodic theory* (Springer, Berlin).

[81] Cosgrove, R. (1996). Consistent evolution with different time slicings in quantum gravity, *Class. Quant. Grav.* **13**, 891–920.

[82] Coussaert, O., Henneaux, M., and van Driel, P. (1995). The asymptotic dynamics of three-dimensional Einstein gravity with a negative cosmological constant, *Class. Quant. Grav.* **12**, 2961–2966.

[83] Creighton, J. D. E. and Mann, R. B. (1994). Temperature, energy, and heat capacity of asymptotically anti-de Sitter black holes, *Phys. Rev.* **D50**, 6394–6403.

[84] Criscuolo, A., Quevedo, H., and Waelbroeck, H. (1997). Quantization of 2+1 gravity on the torus, in *Field theory, integrable systems and symmetries*, ed. F. Khanna and L. Vinet (Les Publications CRM, Montréal).

[85] Crnkovic, C. and Witten, E. (1987). Covariant description of canonical formalism in geometric theories, in *Three hundred years of gravitation*, ed. S. W. Hawking and W. Israel (Cambridge University Press, Cambridge).

[86] DePietri, R. and Rovelli, C. (1996). Geometry, eigenvalues, and scalar product from recoupling theory in loop quantum gravity, *Phys. Rev.* **D54**, 2664–2690.

[87] Deser, S. (1985). On the breakdown of asymptotic Poincaré invariance in $D = 3$ Einstein gravity, *Class. Quant. Grav.* **2**, 489–495.

[88] Deser, S. and Jackiw, R. (1984). Three-dimensional cosmological gravity: dynamics of constant curvature, *Ann. Phys.* **153**, 405–416.

[89] Deser, S. and Jackiw, R. (1988). Classical and quantum scattering on a cone, *Commun. Math. Phys.* **118**, 495–509.

[90] Deser, S., Jackiw, R., and 't Hooft, G. (1984). Three-dimensional Einstein gravity: dynamics of flat space, *Ann. Phys.* **152**, 220–235.

[91] Deser, S., Jackiw, R., and Templeton, S. (1982a). Three-dimensional massive gauge theories, *Phys. Rev. Lett.* **48**, 975–978.

[92] Deser, S., Jackiw, R., and Templeton, S. (1982b). Topologically massive gauge theories, *Ann. Phys.* **140**, 372–411.

[93] Deser, S. and Xiang, X. (1991). Canonical formulations of full nonlinear topologically massive gravity, *Phys. Lett.* **B263**, 39–43.

[94] Deser, S. and Yang, Z. (1990). Is topologically massive gravity renormalizable?, *Class. Quant. Grav.* **7**, 1603–1612.

[95] de Wit, B., Matschull, H.-J., and Nicolai, H. (1993). Physical states in $d = 3$, $N = 2$ supergravity, *Phys. Lett.* **B318**, 115–121.

[96] DeWitt, B. S. (1964). Gravity, a universal regulator? *Phys. Rev. Lett.* **13**, 114–118.

[97] DeWitt, B. S. (1967). Quantum gravity. I. The canonical theory, *Phys. Rev.* **160**, 1113–1148.

[98] D'Hoker, E. and Giddings, S. B. (1987). Unitarity of the closed bosonic Polyakov string, *Nucl. Phys.* **B291**, 90–112.

[99] D'Hoker, E. and Phong, D. H. (1988). The geometry of string perturbation theory, *Rev. Mod. Phys.* **60**, 917–1065.

[100] Dirac, P. A. M. (1950). Generalized Hamiltonian dynamics, *Can. J. Math.* **2**, 129–148.

[101] Dirac, P. A. M. (1951). The Hamiltonian form of field dynamics, *Can. J. Math.* **3**, 1–23.

[102] Dirac, P. A. M. (1958). Generalized Hamiltonian dynamics, *Proc. Roy. Soc. (London)* **A246**, 326–332.

[103] Drumm, T. A. (1992). Fundamental polyhedra for Margulis spacetimes, *Topology* **31**, 677–683.

[104] Drumm, T. A. and Goldman, W. M. (1990). Complete flat Lorentzian 3-manifolds with free fundamental group, *Int. J. Math.* **1**, 149–161.

[105] Eguchi, T., Gilkey, P. B., and Hanson, A. J. (1980). Gravitation, gauge theories and differential geometry, *Physics Reports* **66**, 213–393.

[106] Ezawa, K. (1994a). Classical and quantum evolution of the de Sitter and the anti-de Sitter universes in (2+1)-dimensions, *Phys. Rev.* **D49**, 5211–5226; addendum, *Phys. Rev.* **D50**, 2935–2938.

[107] Ezawa, K. (1994b). Transition amplitude in (2+1)-dimensional Chern–Simons gravity on a torus, *Int. J. Mod. Phys.* **A9**, 4727–4746.

[108] Ezawa, K. (1995). Chern–Simons quantization of (2+1)-anti-de Sitter gravity on a torus, *Class. Quant. Grav.* **12**, 373–391.

[109] Ezawa, K. (1996). Combinatorial solutions to the Hamiltonian constraint in (2+1)-dimensional Ashtekar gravity, *Nucl. Phys.* **B459**, 355–392.

[110] Farkas, H. M. and Kra, I. (1980). *Riemann Surfaces* (Springer, New York).

[111] Fay, J. D. (1977). Fourier coefficients of the resolvent for a Fuchsian group, *J. Reine Angw. Math.* **293**, 143–203.

[112] Franzosi, R. and Guadagnini, E. (1996). Topology and classical geometry in (2+1) gravity, *Class. Quant. Grav.* **13**, 433–460.

[113] Fried, D. and Goldman, W. (1983). Three-dimensional affine crystallographic groups, *Adv. Math.* **47**, 1–49.

[114] Friedman, J. L. and Sorkin, R.D. (1980). Spin 1/2 from gravity, *Phys. Rev. Lett.* **44**, 1100–1103.

[115] Fujii, M. (1990). Hyperbolic 3-manifolds with totally geodesic boundary, *Osaka J. Math.* **27**, 539–553.

[116] Fujiwara, Y. (1993). Geometrical construction of holonomy in a 3-dimensional hyperbolic manifold, *Class. Quant. Grav.* **10**, 219–232.

[117] Fujiwara, Y., Higuchi, S., Hosoya, A., Mishima, T., and Siino, M. (1991). Nucleation of a universe in (2+1)-dimensional gravity with a negative cosmological constant, *Phys. Rev.* **D44**, 1756–1762.

[118] Fujiwara, Y., Higuchi, S., Hosoya, A., Mishima, T., and Siino, M. (1992). Topology change by quantum tunneling in (2+1)-dimensional Einstein gravity, *Prog. Theor. Phys.* **87**, 253–268.

[119] Fujiwara, Y. and Soda, J. (1990). Teichmüller motion of (2+1)-dimensional gravity with the cosmological constant, *Prog. Theor. Phys.* **83**, 733–748.

[120] Gardiner, F. P. (1987). *Teichmüller theory and quadratic differentials* (Wiley, New York).

[121] Gegenberg, J. and Kunstatter, G. (1994). The partition function for topological field theories, *Ann. Phys.* **231**, 270–289.

[122] Geroch, R. (1976). Domain of dependence, *J. Math. Phys.* **11**, 437–449.

[123] Geroch, R. and Horowitz, G. T. (1979). Global structure of spacetimes, in *General relativity, an Einstein centenary survey*, ed. S. W. Hawking and W. Israel (Cambridge University Press, Cambridge).

[124] Gibbons, G. W. and Hartle, J. B. (1990). Real tunneling geometries and the large scale topology of the universe, *Phys. Rev.* **D42**, 2458–2468.

[125] Gibbons, G. W. and Hawking, S. W. (1977). Action integrals and partition functions in quantum gravity, *Phys. Rev.* **D15**, 2752–2756.

[126] Giddings, S. B. (1996). The black hole information paradox, in *Particles, strings and cosmology: proceedings of the Johns Hopkins workshop on current problems in particle theory*, ed. J. Bagger, G. Domokos, A. Falk, and S. Kovesi-Domokos (World Scientific, Singapore).

[127] Giulini, D. (1997). Determination and reduction of large diffeomorphisms, in *Constrained Dynamics and Quantum Gravity 1996*, ed. V. de Alfaro, J. E. Nelson, G. Bandelloni *et al.* (North-Holland, Amsterdam).

[128] Giulini, D. and Louko, J. (1995). Diffeomorphism invariant subspaces in Witten's (2+1) quantum gravity on $\mathbf{R} \times T^2$, *Class. Quant. Grav.* **12**, 2735–2746.

[129] Goldman, W. M. (1984). The symplectic nature of fundamental groups of surfaces, *Adv. Math.* **54**, 200–225.

[130] Goldman, W. M. (1988). Topological components of spaces of representations, *Invent. Math.* **93**, 557–607.

[131] Goldman, W. M. and Kamishima, Y. (1984). The fundamental group of a compact flat Lorentz space form is virtually polycyclic, *J. Diff. Geom.* **19**, 233–240.

[132] Goni, M. A. and Valle, M. A. (1986). Massless fermions and (2+1)-dimensional gravitational effective action, *Phys. Rev.* **D34**, 648–649.

[133] Gonzalez, G. and Pullin, J. (1990). BRST analysis of (2+1) gravity, *Phys. Rev.* **D42**, 3395–3400; erratum, *Phys. Rev.* **D43**, 2749 (1991).

[134] Gott, J. R. (1991). Closed timelike curves produced by pairs of moving cosmic strings: exact solutions, *Phys. Rev. Lett.* **66**, 1126–1129.

[135] Govaerts, J. (1991). *Hamiltonian quantisation and constrained dynamics* (Leuven University Press, Leuven).

[136] Grignani, G. and Nardelli, G. (1992). Gravity and the Poincaré group, *Phys. Rev.* **D45**, 2719–2731.

[137] Grignani, G., Sodano, P., and Scrucca, C. A. (1996). Quantum states of topologically massive electrodynamics and gravity, *J. Phys. A: Math. Gen.* **29**, 3179–3197.

[138] Gross, M. and Varsted, S. (1992). Elementary moves and ergodicity in D-dimensional simplicial quantum gravity, *Nucl. Phys.* **B378**, 367–380.

[139] Hamber, H. W. and Williams, R. M. (1993). Simplicial quantum gravity in three dimensions: analytical and numerical results, *Phys. Rev.* **D47**, 510–532.

[140] Hartle, J. B. and Hawking, S. W. (1983). The wave function of the universe, *Phys. Rev.* **D28**, 2960–2975.

[141] Harvey, W. J. (1977). *Discrete groups and automorphic functions* (Academic Press, London).

[142] Hasslacher, B. and Perry, M. J. (1981). Spin networks are simplicial quantum gravity, *Phys. Lett.* **B103**, 21–24.

[143] Hawking, S. W. (1974). Black hole explosions, *Nature* **248**, 30–31.

[144] Hawking, S. W. (1975). Particle creation by black holes, *Commun. Math. Phys.* **43**, 199–220.

[145] Hawking, S. W. and Ellis, G. F. R. (1973). *The large scale structure of space-time* (Cambridge University Press, Cambridge).

[146] Hempel, J. (1976). *3-Manifolds*, Ann. Math. Studies 86 (Princeton University Press, Princeton).

[147] Henneaux, M. (1984a). Energy–momentum, angular momentum, and supercharge in 2+1 supergravity, *Phys. Rev.* **D29**, 2766–2768.

[148] Henneaux, M. (1984b). Coordinate invariant states for quantum gravity in 2+1 dimensions, *Phys. Lett.* **134B**, 184–186.

[149] Henneaux, M. and Teitelboim, C. (1992). *Quantization of gauge systems* (Princeton University Press, Princeton).

[150] Horne, J. H. and Moore, G. (1994). Chaotic coupling constants, *Nucl. Phys.* **B432**, 109–126.

[151] Horowitz, G. (1996). The origin of black hole entropy in string theory, Santa Barbara preprint UCSBTH-96-07, gr-qc/9604051, to appear in *Proc. of the Pacific Conference on Gravitation and Cosmology (Seoul, Korea)*.

[152] Hosoya, A. (1995). Quantum smearing of spacetime singularity, *Class. Quant. Grav.* **12**, 2967–2975.

[153] Hosoya, A. and Nakao, K. (1990). (2+1)-dimensional pure gravity for an arbitrary closed initial surface, *Class. Quant. Grav.* **7**, 163–176.

[154] Hyun, S., Lee, G. H., and Yee, J. H. (1994). Hawking radiation from a (2+1)-dimensional black hole, *Phys. Lett.* **B322**, 182–187.

[155] Hyun, S., Song, Y.-S., and Yee, J. H. (1995). Hawking radiation of Dirac fields in the (2+1)-dimensional black hole space-time, *Phys. Rev.* **D51**, 1787–1792.

[156] Ichinose, I. and Satoh, Y. (1995). Entropies of scalar fields on three dimensional black holes, *Nucl. Phys.* **B447**, 340–370.

[157] Ionicioiu, R. (1996). Building blocks in Turaev–Viro theory, Cambridge preprint DAMTP-96-94, gr-qc/9611024.

[158] Isham, C. J. (1984). Topological and global aspects of quantum theory, in *Relativity, groups and topology II*, ed. B. S. DeWitt and R. Stora (North-Holland, Amsterdam).

[159] Isham, C. J. (1989). *Modern differential geometry for physicists* (World Scientific, Singapore).

[160] Isham, C. J. (1994). Prima facie questions in quantum gravity, in *Canonical gravity: from classical to quantum*, Lecture Notes in Physics 434, ed. J. Ehlers and H. Friedrich (Springer, New York).

[161] Isham, C. J., Salam, A., and Strathdee, J. (1971). Infinity suppression in gravity-modified quantum electrodynamics, *Phys. Rev.* **D3**, 1805–1817.

[162] Isham, C. J., Salam, A., and Strathdee, J. (1972). Infinity suppression in gravity-modified quantum electrodynamics II, *Phys. Rev.* **D5**, 2548–2565.

[163] Ishibashi, N. (1990). Quantum Legendre transformation and quantum gravity, Santa Barbara preprint UCSBTH-90-72.

[164] Jänich, K. (1984). *Topology* (Springer, New York).

[165] Jackiw, R. (1984). Topological investigations of quantized gauge theories, in *Relativity, groups and topology II*, ed. B. S. DeWitt and R. Stora (North-Holland, Amsterdam).

[166] Kabat, D. and Ortiz, M. E. (1994). Canonical quantization and braid group invariance of (2+1)-dimensional gravity coupled to point particles, *Phys. Rev.* **D49**, 1684–1688.

[167] Kauffman, L. (1992). Map coloring, q-deformed spin networks, and the Turaev–Viro invariant for 3-manifolds, *Int. J. Mod. Phys.* **B6**, 1765–1794.

[168] Keszthelyi, B. and Kleppe, G. (1992). Renormalizability of $D = 3$ topologically massive gravity, *Phys. Lett.* **B281**, 33–35.

[169] Kojima, S. and Miyamoto, Y. (1991). The smallest hyperbolic 3-manifolds with totally geodesic boundary, *J. Diff. Geom.* **34**, 175–192.

[170] Komar, A. (1958). Construction of a complete set of independent observables in the general theory of relativity, *Phys. Rev.* **111**, 1182–1187.

[171] Kuchař, K. (1992). Time and interpretations of quantum gravity, in *General relativity and relativistic astrophysics, proceedings of the 4th Canadian conference*, ed. G. Kunstatter, D. E. Vincent, and J. G. Williams (World Scientific, Singapore).

[172] Kuchař, K. (1993). Canonical quantum gravity, in *General relativity and gravitation 1992*, ed. R. J. Gleiser, C. N. Kozameh, and O. M. Moreschi (Institute of Physics Publishing, Bristol).

[173] Kugo, T. (1990). Renormalizability of 3-dimensional gravity coupled to matter, Kyoto preprint KUNS 1014 HE(TH)90/05.

[174] Landau, L. D. and Lifshitz, E. M. (1977). *Quantum mechanics (non-relativistic theory)* (Pergamon, Oxford).

[175] Lee, J. and Wald, R. M. (1990). Local symmetries and constraints, *J. Math. Phys.* **31**, 725–743.

[176] Leutwyler, H. (1966). A (2+1)-dimensional model for the quantum theory of gravity, *Nuovo Cimento* **42**, 159–178.

[177] Lifschytz, G. and Ortiz, M. (1994). Scalar field quantization on the (2+1)-dimensional black hole background, *Phys. Rev.* **D49**, 1929–1943.

[178] Loll, R. (1995a). Independent loop invariants for (2+1) gravity, *Class. Quant. Grav.* **12**, 1655–1662.

[179] Loll, R. (1995b). Quantum aspects of (2+1) gravity, *J. Math. Phys.* **36**, 6494–6509.

[180] Louko, J. (1995). Witten's 2+1 gravity on **R** × (Klein bottle), *Class. Quant. Grav.* **12**, 2441–2468.

[181] Louko, J. and Marolf, D. (1994). Solution space of (2+1) gravity on **R** × T^2 in Witten's connection formulation, *Class. Quant. Grav.* **11**, 311–330.

[182] Louko, J. and Ruback, P. J. (1991). Spatially flat quantum cosmology, *Class. Quant. Grav.* **8**, 91–121.

[183] Lück, W. (1993). Analytic and topological torsion for manifolds with boundary and symmetry, *J. Diff. Geom.* **37**, 263–322.

[184] Mann, R. B. and Solodukhin, S. N. (1997). Quantum scalar field on three-dimensional (BTZ) black hole instanton: heat kernel, effective action and thermodynamics, *Phys. Rev.* **D55**, 3622–3632.

[185] Marolf, D. (1993). Loop representations for (2+1) gravity on a torus, *Class. Quant. Grav.* **10**, 2625–2648.

[186] Marolf, D. (1995). Refined algebraic quantization: systems with a single constraint, Santa Barbara preprint UCSBTH-95-16, gr-qc/9508015.

[187] Marolf, D., Mourão, J., and Thiemann, T. (1997). The status of diffeomorphism superselection in Euclidean 2+1 gravity, *J. Math. Phys.* **38**, 4730–4740.

[188] Martin, S. P. (1989). Observables in 2+1 dimensional gravity, *Nucl. Phys.* **327**, 178–204.

[189] Martinec, E. (1984). Soluble systems in quantum gravity, *Phys. Rev.* **D30**, 1198–1204.

[190] Matschull, H.-J. (1994). On loop states in quantum gravity and supergravity, *Class. Quant. Grav.* **11**, 2395–2410.

[191] Matschull, H.-J. and Nicolai, H. (1994). Canonical quantum supergravity in three dimensions, *Nucl. Phys.* **B411**, 609–646.

[192] Maunder, C. R. F. (1980). *Algebraic topology* (Cambridge University Press, Cambridge).

[193] Mazur, P. O. (1991). Quantum gravitational measure for three-geometries, *Phys. Lett.* **B262**, 405–410.

[194] McCullough, D. (1986). Mappings of reducible 3-manifolds, in *Geometric and algebraic topology*, Banach Center Publ. **18**, ed. H. Torunczyk, S. Jackowski, and S. Spiez (Polish Scientific Publishers, Warsaw).

[195] Mess, G. (1990). Lorentz spacetimes of constant curvature, Institut des Hautes Etudes Scientifiques preprint IHES/M/90/28.

[196] Milnor, J. W. (1965). *Topology from the differentiable viewpoint* (University Press of Virginia, Charlottesville).

[197] Milnor, J. W. and Stasheff, J. D. (1974). *Characteristic classes*, Ann. Math. Studies 76 (Princeton University Press, Princeton).

[198] Misner, C. W., Thorne, K. S., and Wheeler, J. A. (1973). *Gravitation* (W. H. Freeman, San Francisco).

[199] Mizoguchi, S. and Tada, T. (1992a). Three-dimensional gravity from the Turaev–Viro invariant, *Phys. Rev. Lett.* **68**, 1795–1798.

[200] Mizoguchi, S. and Tada, T. (1992b). 3-dimensional gravity and the Turaev–Viro invariant, *Prog. Theor. Phys. Suppl.* **110**, 207–227.

[201] Mizoguchi, S. and Yamamoto, H. (1994). Stability of renormalizable expansions in three-dimensional gravity, *Phys. Rev.* **D50**, 7351–7362.

[202] Moncrief, V. (1989). Reduction of the Einstein equations in 2+1 dimensions to a Hamiltonian system over Teichmüller space, *J. Math. Phys.* **30**, 2907–2914.

[203] Moore, G. and Nelson, P. (1986). Measure for moduli, *Nucl. Phys.* **B266**, 58–74.

[204] Nash, C. and Sen, S. (1983). *Topology and geometry for physicists* (Academic Press, London).

[205] Nelson, J. E. and Regge, T. (1989). Homotopy groups and 2+1 dimensional quantum gravity, *Nucl. Phys.* **B328**, 190–202.

[206] Nelson, J. E. and Regge, T. (1991a). 2+1 quantum gravity, *Phys. Lett.* **B272**, 213–216.

[207] Nelson, J. E. and Regge, T. (1991b). 2+1 gravity for genus > 1, *Commun. Math. Phys.* **141**, 211–223.

[208] Nelson, J. E. and Regge, T. (1992). 2+1 quantum gravity for high genus, *Class. Quant. Grav.* **9**, 187–196.

[209] Nelson, J. E., Regge, T., and Zertuche, F. (1990). Homotopy groups and 2+1 dimensional quantum de Sitter gravity, *Nucl. Phys.* **B339**, 516–532.

[210] Neumann, W. D. and Reid, A. W. (1991a). Amalgamation and the invariant trace field of a Kleinian group, *Math. Proc. Camb. Philos. Soc.* **109**, 509–515.

[211] Neumann, W. D. and Reid, A. W. (1991b). Rigidity of cusps in deformations of hyperbolic 3-orbifolds, *Math. Ann.* **295**, 223–237.

[212] Newbury, P. and Unruh, W. G. (1994). Solution to (2+1) gravity in dreibein formalism, *Int. J. Mod. Phys.* **D3**, 131–138.

[213] Oda, I. and Yahikozawa, S. (1994). Effective actions of (2+1)-dimensional gravity and BF theory, *Class. Quant. Grav.* **11**, 2653–2666.

[214] Okamura, T. and Ishihara, H. (1992). Perturbation of the higher genus spatial surface in (2+1)-dimensional gravity, *Phys. Rev.* **D46**, 572–576.

[215] Okamura, T. and Ishihara, H. (1993). Perturbation of higher genus surfaces in (2+1)-dimensional gravity with a cosmological constant, *Phys. Rev.* **D47**, 1706–1708.

[216] Ooguri, H. (1992). Partition functions and topology-changing amplitudes in the three-dimensional lattice gravity of Ponzano and Regge, *Nucl. Phys.* **B382**, 276–304.

[217] Ooguri, H. (1993). Schwinger–Dyson equation in three-dimensional simplicial quantum gravity, *Prog. Theor. Phys.* **89**, 1–22.

[218] Ooguri, H. and Sasakura, N. (1991). Discrete and continuum approaches to 3-dimensional quantum gravity, *Mod. Phys. Lett.* **A6**, 3591–3600.

[219] Peldan, P. (1996a). Large diffeomorphisms in (2+1) quantum gravity on the torus, *Phys. Rev.* **D53**, 3147–3155.

[220] Peldan, P. (1996b). A modular invariant quantum theory from the connection formulation of (2+1) gravity on the torus, *Class. Quant. Grav.* **13**, 221–224.

[221] Ponzano, G. and Regge, T. (1968). Semiclassical limit of Racah coefficients, in *Spectroscopic and group theoretical methods in physics*, ed. F. Bloch, S. G. Cohen, A. De-Shalit, S. Sambursky, and I. Talmi (North-Holland, Amsterdam).

[222] Puzio, R. (1994a). On the square root of the Laplace–Beltrami operator as a Hamiltonian, *Class. Quant. Grav.* **11**, 609–620.

[223] Puzio, R. S. (1994b). The Gauss map and (2+1) gravity, *Class. Quant. Grav.* **11**, 2667–2676.

[224] Rama, S. K. and Sen, S. (1992). 3D manifolds, graph invariants, and Chern–Simons theory, *Mod. Phys. Lett.* **A7**, 2065–2076.

[225] Rankin, R. A. (1977). *Modular forms and functions* (Cambridge University Press, Cambridge).

[226] Rankin, R. A. (1984). *Modular forms* (Halsted Press, New York).

[227] Ray, D. B. and Singer, I. M. (1971). R-Torsion and the Laplacian on Riemannian manifolds, *Adv. Math.* **7**, 145–210.

[228] Regge, T. (1961). General relativity without coordinates, *Nuovo Cimento* **19**, 558–571.

[229] Regge, T. and Teitelboim, C. (1974). Role of surface integrals in the Hamiltonian formulation of general relativity, *Ann. Phys.* **88**, 286–318.

[230] Rovelli, C. (1990). Quantum mechanics without time: a model, *Phys. Rev.* **D42**, 2638–2646.

[231] Rovelli, C. (1991). Time in quantum gravity: physics beyond the Schrödinger regime, *Phys. Rev.* **D43**, 442–456.

[232] Rovelli, C. (1993). The basis of the Ponzano–Regge–Turaev–Viro–Ooguri quantum gravity model is the loop representation basis, *Phys. Rev.* **D48**, 2702–2707.

[233] Sa, P. M., Kleber, A., and Lemos, J. P. S. (1996). Black holes in three-dimensional dilaton gravity theories, *Class. Quant. Grav.* **13**, 125–137.

[234] Sasakura, N. (1992). Exact three-dimensional lattice gravities, *Prog. Theor. Phys. Suppl.* **110**, 191–205.

[235] Schleich, K. and Witt, D. M. (1993). Generalized sums over histories for quantum gravity II: simplicial conifolds, *Nucl. Phys.* **B402**, 469–528.

[236] Schoen, R. (1984). Conformal deformation of a Riemannian metric to constant scalar curvature, *J. Diff. Geom.* **20**, 479–495.

[237] Schwarz, A. S. (1978). The partition function of degenerate quadratic functional and Ray–Singer invariants, *Lett. Math. Phys.* **2**, 247–252.

[238] Segal, G. B. (1989). Conformal field theory, in *International Congress on Mathematical Physics*, ed. B. Simon, A. Truman, and I. M. Davies (A. Hilger, Bristol).

[239] Seriu, M. (1997). Partition function for (2+1)-dimensional Einstein gravity, *Phys. Rev.* **D55**, 781–790.

[240] Shiraishi, K. and Maki, T. (1994a). Vacuum polarization around a three-dimensional black hole, *Class. Quant. Grav.* **11**, 695–699.

[241] Shiraishi, K. and Maki, T. (1994b). Quantum fluctuation of stress tensor and black holes in three dimensions, *Phys. Rev.* **D49**, 5286–5294.

[242] Sorkin, R. D. (1986). Non-time-orientable Lorentzian cobordism allows for pair creation, *Int. J. Theor. Phys.* **25**, 877–881.

[243] Staruszkiewicz, A. (1963). Gravitation theory in three-dimensional space, *Acta Physica Polonica* **6**, 735–740.

[244] Steif, A. (1994). The quantum stress tensor in the three-dimensional black hole, *Phys. Rev.* **D49**, 585–589.

[245] Steif, A. (1996). Supergeometry of three-dimensional black holes, *Phys. Rev.* **D53**, 5521–5526.

[246] Strominger, A. and Vafa, C. (1996). Microscopic origin of the Bekenstein–Hawking entropy, *Phys. Lett.* **B379**, 99–104.

[247] 't Hooft, G. (1988). Non-perturbative two particle scattering amplitudes in 2+1 dimensional gravity, *Commun. Math. Phys.* **117**, 685–700.

[248] 't Hooft, G. (1992a). Causality in (2+1)-dimensional gravity, *Class. Quant. Grav.* **9**, 1335–1348.

[249] 't Hooft, G. (1992b). Classical N-particle cosmology in 2+1 dimensions, *Class. Quant. Grav.* **10**, S79–S91.

[250] 't Hooft, G. (1993a). The evolution of gravitating point particles in 2+1 dimensions, *Class. Quant. Grav.* **10**, 1023–1038.

[251] 't Hooft, G. (1993b). Canonical quantization of gravitating point particles in 2+1 dimensions, *Class. Quant. Grav.* **10**, 1653–1664.

[252] 't Hooft, G. (1996). Quantization of point particles in (2+1)-dimensional gravity and spacetime discreteness, *Class. Quant. Grav.* **13**, 1023–1039.

[253] Teitelboim, C. (1973). How commutators of constraints reflect the spacetime structure, *Ann. Phys.* **79**, 542–557.

[254] Teitelboim, C. (1983). Causality versus gauge invariance in quantum gravity and supergravity, *Phys. Rev. Lett.* **50**, 705–708.

[255] Terras, A. (1985). *Harmonic analysis on symmetric spaces and applications I* (Springer, New York).

[256] Thurston, W. P. (1979). The geometry and topology of three-manifolds, Princeton lecture notes.

[257] Thurston, W. P. (1982). Three dimensional manifolds, Kleinian groups, and hyperbolic geometry, *Bull. Amer. Math. Soc.* **6**, 357–381.

[258] Turaev, V. G. (1991). Quantum invariants of 3-manifolds and a glimpse of shadow topology, *C. R. Acad. Sci. Paris* **313**, 395–398.

[259] Turaev, V. G. (1994). *Quantum invariants of knots and 3-manifolds* (Walter de Gruyter, Berlin).

[260] Turaev, V. G. and Viro, O. Y. (1992). State sum invariants of 3-manifolds and quantum $6j$-symbols, *Topology* **31**, 865–902.

[261] Unruh, W. G. and Wald, R. M. (1989). Time and the interpretation of quantum gravity, *Phys. Rev.* **D40**, 2598–2614.

[262] van der Bij, J. J., Pisarski, R. D., and Rao, S. (1986). Topological mass term for gravity induced by matter, *Phys. Lett.* **B179**, 87–91.

[263] Van Hove, L. (1951). Sur le probléme des relations entre les transformations unitaires de le mécanique quantique et les transformations canoniques de la mécanique classique, *Acad. Roy. Belgique Bull. Cl. Sci.* **37**, 610–620.

[264] Verlinde, H. (1990). Conformal field theory, 2-dimensional quantum gravity and quantization of Teichmüller space, *Nucl. Phys.* **B337**, 652–680.

[265] Vuorio, I. (1986). Parity violation and the effective gravitational action in three dimensions, *Phys. Lett.* **B175**, 176–178.

[266] Waelbroeck, H. (1990a). 2+1 lattice gravity, *Class. Quant. Grav.* **7**, 751–769.

[267] Waelbroeck, H. (1990b). Time-dependent solutions of 2+1 gravity, *Phys. Rev. Lett.* **64**, 2222–2225.

[268] Waelbroeck, H. (1991). Solving the time-evolution problem in 2+1 gravity, *Nucl. Phys.* **B364**, 475–494.

[269] Waelbroeck, H. (1994). Canonical quantization of (2+1)-dimensional gravity, *Phys. Rev.* **D50**, 4982–4992.

[270] Waelbroeck, H. and Zapata, J. A. (1993). Translation symmetry in 2+1 Regge calculus, *Class. Quant. Grav.* **10**, 1923–1932.

[271] Waelbroeck, H. and Zapata, J. A. (1996). (2+1) covariant lattice theory and 't Hooft's formulation, *Class. Quant. Grav.* **13**, 1761–1768.

[272] Waelbroeck, H. and Zertuche, F. (1994). Homotopy invariants and time evolution in (2+1)-dimensional gravity, *Phys. Rev.* **D50**, 4966–4981.

[273] Wagoner, R. V. (1970). Scalar–tensor theory and gravitational waves, *Phys. Rev.* **D1**, 3209–3216.

[274] Wald, R. M. (1984). *General Relativity* (University of Chicago Press, Chicago).

[275] Wald, R. M. (1993). Black hole entropy is a Noether charge, *Phys. Rev.* **D48**, R3427–R3431.

[276] Wald, R. M. (1994). *Quantum field theory in curved spacetime and black hole thermodynamics* (University of Chicago Press, Chicago).

[277] Weeks, J. R. (1985). *The shape of space* (Marcel Dekker, New York).

[278] Weinberg, S. (1995). *The quantum theory of fields* (Cambridge University Press, Cambridge).

[279] Welling, M. (1996). Gravity in (2+1)-dimensions as a Riemann–Hilbert problem, *Class. Quant. Grav.* **13**, 653–680.

[280] Welling, M. (1997). The torus universe in the polygon approach to (2+1)-dimensional gravity, *Class. Quant. Grav.* **14**, 929–943.

[281] Welling, M. and Bijlsma, M. (1996). The Pauli–Lubanski scalar in the polygon approach to (2+1)-dimensional gravity *Class. Quant. Grav.* **13**, 1769–1774.

[282] Wen, X. G. (1990a). Electrodynamic properties of gapless excitations in the fractional quantum Hall states, *Phys. Rev. Lett.* **64**, 2206–2209.

[283] Wen, X. G. (1990b). Non-Abelian statistics in the fractional quantum Hall states, *Phys. Rev. Lett.* **66**, 802–805.

[284] Wen, X. G. (1992). Theory of the edge states in fractional quantum Hall effects, *Int. J. Mod. Phys* **B6**, 1711–1762.

[285] Wheeler, J. A, (1968). Superspace and the nature of quantum geometrodynamics, in *Battelle rencontres, 1967 lectures in mathematics and physics,* ed. C. DeWitt and J. A. Wheeler (W. A. Benjamin, New York).

[286] Williams, R. M. (1992). Discrete quantum gravity: the Regge calculus approach, *Int. J. Mod. Phys.* **B6**, 2097–2108.

[287] Witten, E. (1988). 2+1 dimensional gravity as an exactly soluble system, *Nucl. Phys.* **B311**, 46–78.

[288] Witten, E. (1989a). Topology-changing amplitudes in 2+1 dimensional gravity, *Nucl. Phys.* **B323**, 113–140.

[289] Witten, E. (1989b). Quantum field theory and the Jones polynomial, *Commun. Math. Phys.* **121**, 351–399.

[290] Witten, E. (1991). Quantization of Chern–Simons gauge theory with complex gauge group, *Commun. Math. Phys.* **137**, 29–66.

[291] Witten, E. (1992). On holomorphic factorization of WZW and coset models, *Commun. Math. Phys.* **144**, 189–212.

[292] Witten, E. (1995). Is supersymmetry really broken? *Int. J. Mod. Phys.* **A10**, 1247–1248.

[293] Woodard, R. P. (1993). Enforcing the Wheeler–DeWitt constraint the easy way, *Class. Quant. Grav.* **10**, 483–496.

[294] Wu, S. (1991). Topological quantum field theories on manifolds with a boundary, *Commun. Math. Phys.* **136**, 157–168.

[295] Yamabe, H. (1960). On a deformation of Riemannian structure on compact manifolds, *Osaka Math. J.* **12**, 21–37.

[296] Yodzis, P. (1972). Lorentz cobordism, *Commun. Math. Phys.* **27**, 39–52.

[297] York, J. W. (1972). Role of conformal three-geometry in the dynamics of gravitation, *Phys. Rev. Lett.* **28**, 1082–1085.

[298] York, J. W. (1974). Covariant decompositions of symmetric tensors in the theory of gravitation, *Ann. Inst. Henri Poincaré* **A21**, 319–332.

[299] Zwiebach, B. (1993). Closed string field theory: quantum action and the Batalin–Vilkovisky master equation, *Nucl. Phys.* **B390**, 33–152.

Index

action
    Chern–Simons, 28
    dynamical triangulation, 192
    Einstein–Hilbert, 3, 14
        canonical form, 14
        conformal properties, 150
    Euclidean, 166, 181
    first-order, 26, 34, 152
    reduced phase space, 19, 89
    Regge, 172, 173
    scalar field, 30, 199
ADM formalism, 11–15
    and classical solutions, 38–41
    canonical momenta, 14
adS, *see* anti-de Sitter space
Aharonov–Bohm phases, 28, 60, 249
analytic torsion, *see* torsion, Ray–
    Singer
angular momentum, 25, 43–5, 49, 196,
    198
anomaly, 154, 156
anti-de Sitter space, 49, 69, 197
    boundary conditions, 199
    scale, 200
    upper half-space model, 50
Ashtekar variables, 112–14
automorphic forms, 93, 95, 109–11
    inner product, 96
    weight, 95

background field method, 153, 213
BF theories, 155

big bang/big crunch, 53
binor calculus, 180
black hole, 46–50, 194–211, 215
    angular momentum, 49, 198
    as quotient of adS, 49–50, 76,
        196–8
    degrees of freedom, 209–11
    entropy, 195–6, 202, 210
    Euclidean, 202–9
        as quotient of H^3, 204–6
        holonomies, 204
        periodicity, 205–6
    final fate, 216
    holonomies, 76, 198, 204
    horizon, as boundary, 209–10
    horizons, 46–7
    Kruskal coordinates, 47–9, 197
    mass, 49, 198
    Penrose diagram, 48
    statistical mechanics, 209–11
    supersymmetric, 198
    surface gravity, 195
    temperature, 195, 200–1
    thermodynamics, 195–6, 208
    two-point function, 198
        periodicity, 200
Bogoliubov coefficients, 201
bone (in Regge calculus), 172
boundary conditions
    absolute, 158
    Dirichlet, 30–1, 158
    in path integral, 157–8

267